Foreword

BACKGROUND

In the UK, knowledge and experience in dealing with contaminated land have been gained mainly through reclaiming contaminated derelict sites. Information on various aspects of managing contaminated land is available from a number of sources; some, particularly that concerned with treatment technologies, originates from outside the UK. This Volume is one of twelve that attempt to collate this information and the underlying experience into a comprehensive reference and guidance document.

PURPOSE AND SCOPE

The objective is to assist those involved in remediating contaminated land to select and implement effective, economic and safe solutions.

UK practice and procedures, and international experience in the use of techniques that have seen only limited application in the UK to date, have been reviewed during the preparation of the individual Volumes.

The report:

- reflects 'good practice' not only in the context of current policy and regulations but also as they are expected to develop in the foreseeable future.

- is applicable to most contaminated land situations, including those, such as highways and operational industrial sites, which do not conform with the classic 'redevelopment' scenario.

- is designed to meet the needs of a wide range of potential users including project and development managers, consultants and contractors acting on behalf of public and private development agencies, other clients of the construction industry, central and local government, and other regulatory authorities.

Although intended to be as comprehensive as possible, in terms of addressing the full range of hazards that may be encountered, certain subjects are *not* specifically covered:

- the investigation of, and protection from, methane and associated gases - these are the subject of parallel CIRIA studies

- the decontamination and remediation of radioactive contamination

- the removal and/or treatment of asbestos

- physical demolition

- grant regimes and other financial assistance.

Safety is an important and consistent theme within each Volume. Individual sections refer as appropriate to the guidance available from the Health and Safety Executive, and to complementary CIRIA reports giving guidance on safe working practices on contaminated sites and on the use of substances hazardous to health in construction.

STRUCTURE AND CONTENT

The report is in twelve Volumes. Collectively they cover all aspects of remediating contaminated land. Each Volume deals with a specific subject and promotes a systematic approach to the management and remediation of contaminated land (see Table).

Table *Outline structure and contents*

	Volume and title	Outline content
I	INTRODUCTION AND GUIDE	Aims, scope, contents list and brief summaries
II	DECOMMISSIONING, DECONTAMINATION AND DEMOLITION	Issues to be addressed and guidance on procedures
III	SITE INVESTIGATION AND ASSESSMENT	Issues to be addressed and guidance on procedures
IV	CLASSIFICATION AND SELECTION OF REMEDIAL METHODS	Classification and selection of appropriate methods and strategies
V	EXCAVATION AND DISPOSAL	Description and evaluation of methods and guidance on procedures
VI	CONTAINMENT AND HYDRAULIC MEASURES	Description and evaluation of methods and guidance on procedures
VII	EX-SITU REMEDIAL METHODS FOR SOILS, SLUDGES AND SEDIMENTS	Description and evaluation of methods and guidance on procedures
VIII	EX-SITU REMEDIAL METHODS FOR CONTAMINATED GROUNDWATER AND OTHER LIQUIDS	Description and evaluation of methods and guidance on procedures
IX	IN-SITU METHODS OF REMEDIATION	Description and evaluation of methods and guidance on procedures
X	SPECIAL SITUATIONS	Information and guidance on procedures
XI	PLANNING AND MANAGEMENT	Issues to be addressed and guidance on procedures
XII	POLICY AND LEGISLATION	Information on policy, administration and legal frameworks in UK and overseas

USERS

The report is intended for:

- Owners of contaminated sites proposing to take remedial action as a prelude to sale or disposal of land and property; as a precursor to redevelopment; as part of a corporate environmental management programme; or because action is needed to avert a public health and/or environmental threat.

- Non-specialist managers who, faced with redeveloping or remediating contaminated sites, need information and guidance for procurement and project management purposes.

- Non-specialist civil-engineering, architectural or construction advisers providing design, supervision and inspection services in collaboration with specialist advisers.

Special P   1995

Remedial Treatment for Contaminated Land

Volume III: Site investigation and assessment

CONSTRUCTION INDUSTRY RESEARCH AND INFORMATION ASSOCIATION
6 Storey's Gate London SW1P 3AU
E-mail switchboard @ ciria.org.uk
Tel: (0171) 222 8891   Fax: (0171) 222 1708

Summary

Volume III describes the various phases and stages of investigating and assessing a site, and the issues to be addressed throughout. The necessity for investigation and assessment to be set within a risk management framework is highlighted. The basics of risk assessment, the use of guidelines and standards for assessment, and the relationship with subsequent risk reduction and control strategies are described. Emphasis is placed on the importance of setting clear, quantified objectives.

Remedial treatment for contaminated land
Volume III — Site investigation and assessment
Construction Industry Research and Information Association
Special Publication 103, 1995

Keywords
Site investigation, site assessment, risk assessment, risk management

Reader Interest
Contractors, Consulting Engineers, Developers, Local Authorities

All rights reserved. No part of this publication may be reproduced or transmitted in any form or by any means, including photocopying and recording, without the written permission of the copyright holder, application for which should be addressed to the publisher. Such written permission must also be obtained before any part of this publication is stored in a retrieval system of any nature.

© CIRIA 1995

ISBN Nos:

Volume		ISBN
Volume I	(SP101)	086017 396 8
Volume II	(SP102)	086017 397 6
Volume III	(SP103)	086017 398 4
Volume IV	(SP104)	086017 399 2
Volume V	(SP105)	086017 400 X
Volume VI	(SP106)	086017 401 8
Volume VII	(SP107)	086017 402 6
Volume VIII	(SP108)	086017 403 4
Volume IX	(SP109)	086017 404 2
Volume X	(SP110)	086017 405 0
Volume XI	(SP111)	086017 406 9
Volume XII	(SP112)	086017 407 7
Set of twelve volumes		086017 408 5

CLASSIFICATION	
AVAILABILITY	Unrestricted
CONTENT	Guidance based on best current practice
STATUS	Committee guided
USER	Non-specialist managers

Published by CIRIA, 6 Storey's Gate, Westminster, London SW1P 3AU

- Contracting organisations providing groundworks, drilling, surveying, landscape, laboratory analysis, waste management services, etc. to remediation projects.
- Regulatory bodies having responsibility for public and occupational health and safety, and protection of the environment, at all stages of managing a contaminated site.

It is assumed that users will have a basic understanding of the nature of the problems of land contamination. A number of introductory texts [1,2,3] are available which explain:

- the origins and evolution of contaminated land
- the main types of hazards which are likely to encountered on contaminated sites
- the principal potential impacts associated with the presence of contamination.

Remediation of contaminated land is a specialist activity and the management of some forms of contamination requires a high level of specialist input. For the specialist, information is provided on remedial methods which are not, so far, routinely used in the UK.

RELATIONSHIPS BETWEEN VOLUMES

The report is intended to be used as a single source of information and guidance on the assessment and remediation of contaminated sites. Although each Volume is self-contained to the extent that it covers the principal issues and procedures relevant to the subject area, reference to other Volumes may be necessary for more detailed information and discussion on specific aspects. Extensive cross referencing between the various Sections and Volumes is provided to help users locate this more detailed information where necessary.

Certain Volumes, for example those dealing with decommissioning, site investigation, selection procedures or remedial methods, provide factual and guidance material of direct relevance to particular stages of remediation. Others, such as those concerned with policy and legislation, or planning and management issues, are broader in scope and support the technical content of the other Volumes.

PROFESSIONAL ADVICE

This report is not a substitute for professional advice, which will be required in many of the situations encountered. Moreover, because of the current lack of information in some subject areas, it is difficult to provide definitive guidance on all aspects, in all cases.

Practical guidance is offered based on a comprehensive review of the current 'state of the art' and a consensus view of good practice, although a number of areas of uncertainty are identified including:

- the short and long-term impacts on human health, and the environment, associated with the presence of contaminants in soils and groundwater
- the development status of some remedial methods, and the lack of field demonstration and experience in their use
- the critical role played by case law in the interpretation of legal requirements which means that experienced and up-to-date legal advice should always be sought, particularly on contentious issues or where uncertainties arise in connection with specific sites.

REFERENCES

1. LEACH, B.A. and GOODGER, H.K. *Building on Derelict Land.* Special Publication 78, CIRIA (London) 1991

2. *Recycling Derelict Land.* Fleming G. (ed.) Thomas Telford (London), 1991

3. *Contaminated Land: Problems and Solutions.* Cairney T., (ed.) Blackie Academic and Professional (Glasgow), 1993

Acknowledgements

This report has been prepared on behalf of CIRIA by:

Mary R Harris	ECOTEC Research and Consulting Ltd (now with Clayton Environmental Consultants Ltd)
Susan M Herbert	TBV Science (formerly with DHV (UK) Ltd)
Michael A Smith	Clayton Environmental Consultants Ltd

Advice on legal aspects was provided by Kathy Mylrea of Simmons and Simmons.

The project was guided and the report prepared with the help and guidance of a Project Steering Group, comprising:

Mr K J Potter (Chairman)	ICI Engineering
Dr P Bardos	Nottingham Trent University (previously at Warren Spring Laboratory)
Mr K W Brierley	British Nuclear Fuels Ltd
Dr T Cairney	W A Fairhurst & Partners
Ms J Denner	Department of the Environment
Dr H E Evans	POWERGEN
Mr R Harris	NRA Severn-Trent Region
Mr M James	Land Restoration Systems
Mr P Kirby	Trafford Park Development Corporation
Mr I Loveday	Dames and Moore International
Professor J D Mather	University of London Royal Holloway and Bedford New College
Dr S Munro	British Gas Plc
Mr S Redfearn	The BOC Foundation for the Environment
Dr J F Rees	Celtic Technologies Ltd
Mr J Thompson	Sir Owen Williams and Partners
Mr J S Watson	Scottish Enterprise
Dr P Wood	AEA Technology (previously at Warren Spring Laboratory)

CIRIA's Research Manager for the project was Dr S T Johnson.

The project was financially supported by British Gas Plc, British Nuclear Fuels Ltd, Dames and Moore International, Department of the Environment, Highlands and Islands Enterprise, ICI Engineering, Land Restoration Systems, NRA Severn-Trent Region, POWERGEN, Rohm and Haas (UK) Ltd, Scottish Enterprise, Sir Owen Williams and Partners, Southern Water Plc, The BOC Foundation for the Environment, and Trafford Park Development Corporation.

The authors wish to acknowledge the help of all those who provided detailed information and comment on this Volume: in particular, Judith Petts of Loughborough University (Sections 5 and 6): Glenn W Suter of Oak Ridge National Laboratory, USA (Appendix 12); Simon Pollard (Aspinwall and Co.), Lance Traves (Clayton Environmental Consultants Inc., USA) and John Emery (John Emery Geotechnical Engineering, Canada).

Thanks are also due to Ms Liz Scott of ECOTEC Research and Consulting Ltd for her secretarial support throughout the project.

Contents

List of Figures x
List of Boxes xii
List of Tables xiii

1 INTRODUCTION 1
  1.1 A Risk management approach 1
  1.2 The purpose of investigation 2
  1.3 Types of investigation 5
  References 6

2 PLANNING THE INVESTIGATION 8
  2.1 Scope 8
  2.2 Setting objectives 8
  2.3 Phasing of the investigation 11
  2.4 Preliminary investigation 13
  2.5 Exploratory investigation 18
  2.6 Detailed investigation 19
  2.7 Integration of investigation procedures 21
  2.8 Reporting 23
  2.9 Ecological assessment 23
  2.10 Post-closure survey 24
  References 25

3 IMPLEMENTATION OF SITE INVESTIGATIONS 26
  3.1 Project management 26
  3.2 Legal aspects 27
  3.3 Procurement 27
  3.4 Specification 29
  3.5 Selection of appropriate specialists 30
  3.6 Quality management 32
  3.7 Health and safety 37
  3.8 Environmental protection requirements 40
  3.9 Long-term sampling and off site works 40
  References 40

4 SAMPLING AND TESTING 44
  4.1 Developing a strategy 44
  4.2 Sampling strategies 45
  4.3 Sampling of soils and similar materials 48
  4.4 Sampling groundwater 60
  4.5 Sampling surface water 66
  4.6 Analytical and testing strategies 69
  References 74

5 RISK ASSESSMENT 77
  5.1 Introduction 77
  5.2 Objectives and scope 77
  5.3 Concepts and definitions 80

5.4 Information requirements 86
5.5 Conducting a site-specific risk assessment 91
5.6 Use of models in risk assessment 97
5.7 Uses of risk assessment 99
5.8 Communication of risks 103
References 104

6 GUIDELINES AND STANDARDS 108
6.1 Scope 108
6.2 Types of guidelines and standards 110
6.3 The use of guidelines and standards for assessment 112
6.4 The use of guidelines and standards for remediation 119
6.5 Dealing with variability 119
6.6 UK guidelines and standards 120
References 126

7 INVESTIGATION FOR COMPLIANCE AND PERFORMANCE 129
7.1 Scope 129
7.2 Excavation 130
7.3 Performance of ex-situ process-based methods 131
7.4 Remediation of groundwater through ex-situ treatment (pump and treat) 136
7.5 Performance of in-situ processes 138
7.6 Monitoring containment methods 145
7.7 Evaluation of remedial methods 150
References 151

Appendix 1 Guidance documents on site investigation 154
Appendix 2 Information sources for desk study 165
Appendix 3 Possible hypotheses on the distribution of contaminants 167
Appendix 4 Guidance on specialists 169
Appendix 5 Key elements in the NAMAS accreditation scheme .. 171
Appendix 6 Investigation techniques 173
Appendix 7 Analytical and testing strategies and methods ... 188
Appendix 8 Important concepts and terms for risk assessment .. 201
Appendix 9 Quantifying human health risks 208
Appendix 10 Examples of modelling in risk assessment 211
Appendix 11 International guidelines and standards 214
Appendix 12 Ecological risk assessment 220

List of Figures

Figure 1.1 *Phased investigations for risk assessment and remedy selection* 4
Figure 4.1 *Three sampling designs (A) Regular (square) grid (B) Stratified random (C) Simple random* 51
Figure 4.2 *Examples of Stratified Random Sampling (16 Sampling locations)* 52
Figure 4.3 *Herringbone sampling design* 52
Figure 4.4 *Relationship between a regular sampling grid and a typical modern housing development* 53
Figure 4.5 *Performance of four sampling designs for detecting a circular target occupying 5% of total site area* 54

Figure 4.6	*Performance of square grid and herringbone sampling designs for detecting targets of various shapes. Relative size of each target is 5% of total site area*	55
Figure 4.7	*Number of sampling locations needed to ensure 0.95 probability of success in hitting targets of different relative sizes*	56
Figure 4.8	*Performance of four designs for detecting an elliptical target (aspect ratio 4:1) as a function of orientation. Relative size of target is 5% (top) 2% (middle) and 1% (bottom) of total site area. Number of sampling locations is 30*	57
Figure 4.9	*Variation of contamination with depth and relationship to materials sampled for trial pits C and D*	58
Figure 4.10	*Variation of contamination along length of Trench A*	59
Figure 4.11	*Variation of contamination along length of Trench B*	60
Figure 4.12	*Mounding of water within a landfill causes modification of local groundwater flow patterns*	61
Figure 4.13	*Subsurface distribution of dense non-aqueous phase liquid (DNAPL)*	62
Figure 4.14	*Examples of judgemental sampling (top), systematic sampling (centre), and random sampling approaches (bottom)*	68
Figure 5.1	*Process for a simple site specific risk assessment*	79
Figure 5.2	*Example of possible exposure pathways*	83
Figure 5.3	*Hazard assessment using generic guidelines and other criteria (single contaminant only)*	92
Figure 5.4	*Potential exposure pathways at site (see Box 5.9)*	95
Figure 7.1	*Contamination increases after remediation stops: contaminant concentrations in groundwater may rebound when pump-and-treat operations cease because of residual contaminants trapped in capillaries (see Figure 7.2).*	137
Figure 7.2	*Limitation of effectiveness of flushing action due to contaminants trapped by capillary action*	137
Figure 7.3	*Reduction of residual contaminant mass by pulsed pumping*	138
Figure 7.4	*Zone of residual contamination caused by pumping: pumping creates a zone of depression to trap light organics such as petroleum for removal but also leaves residues below the water table*	139
Figure 7.5	*Cadmium concentration in pumped percolate from in-situ soil washing operation (see Box 7.5)*	142
Figure 7.6	*Cumulative quantity of cadmium removed and pH of the percolate from in-situ soil washing operation (see Box 7.5)*	143
Figure 7.7	*Cumulative amounts of petrol removed during soil vapour extraction: relative contributions of vaporisation and biodegradation (see Box 7.6)*	144
Figure 7.8	*Cumulative amounts of toluene withdrawn in soil vapour extraction system (see Box 7.6)*	145
Figure 7.9	*Concentration of toluene in withdrawn soil vapour (see Box 7.6)*	145
Figure 7.10	*Monitoring system for cover systems.*	148
Figure 7.11	*Testing cement-bentonite barrier wall*	150
Figure A8.1	*Flow chart for fate and transport assessments: Surface water and sediment*	205
Figure A8.2	*Flow chart for fate and transport assessments: Soils and groundwaters*	206
Figure A8.3	*Flow chart for fate and transport assessments: Atmosphere*	207
Figure A10.1	*Effects of adsorption and biodegradation on groundwater modelling results*	211
Figure A12.1	*Relationship between health and environmental evaluations/ecological assessments*	223
Figure A12.2	*Framework for ecological risk assessment*	224
Figure A12.3	*River water quality: flow-diagram of induced effects following exposure to toxic pollutants*	226
Figure A12.4	*Diagram of risk characterisation in ecological epidemiology*	240

List of Boxes

Box 1.1	A risk-based approach to contaminated land	1
Box 1.2	Investigation for construction purposes	6
Box 2.1	Geotechnical parameters to be addressed by the investigation	11
Box 2.2	Output from the preliminary investigation	14
Box 2.3	Site reconnaissance activities	17
Box 2.4	Soil sample types	19
Box 2.5	Benefits of integrated investigations for risk assessment	21
Box 2.6	Typical content of a site investigation report	24
Box 2.7	Presentation of the report	25
Box 3.1	Selection of appropriate specialists	32
Box 3.2	Key aspects of site investigation which should be subject to performance monitoring	33
Box 3.3	Safety checklist	38
Box 3.4	Examples of environmental protection measures during site investigation	40
Box 4.1	Typical hypotheses	45
Box 4.2	Proving a site is uncontaminated	46
Box 4.3	Character and role of sampling personnel	50
Box 4.4	DD175:1988 Recommendations for sampling	50
Box 4.5	Identifying an area of contamination with 95% confidence	54
Box 4.6	Purging	64
Box 4.7	Sampling groundwater	64
Box 4.8	Environmental analysis	70
Box 4.9	Typical analytical package for soils	71
Box 4.10	Typical gasworks contaminants	72
Box 5.1	Concepts and terms used in risk assessment	81
Box 5.2	Examples of potential human health risks in contaminated land applications	82
Box 5.3	Assumptions used to assess risks associated with presence of dioxin in soils	86
Box 5.4	Examples of uncertainties and assumptions used to estimate human health risks	86
Box 5.5	Information needed to assess risks presented by a contaminated site to potable groundwater source	89
Box 5.6	Example of an exposure uptake equation for ingestion of contaminated soil	94
Box 5.7	Toxicity assessment in the United States	94
Box 5.8	Acceptability of hazards and risks in the United States	97
Box 5.9	Case study example of a quantified risk estimation	98
Box 5.10	Example of derivation of a site-specific remediation value	100
Box 6.1	Important terms and concepts	112
Box 6.2	Contamination and pollution	113
Box 6.3	Background and reference values for soils	114
Box 6.4	Environmental quality standards and objectives	114
Box 6.5	Multi-functionality of soils and waters	117
Box 6.6	Black, grey and red list substances	123
Box 6.7	Occupational exposure limits	124
Box 7.1	Monitoring versus evaluation	130
Box 7.2	Assessment of the effectiveness of a thermal treatment process	133
Box 7.3	Sampling and assessment procedures for the case study described in Box 7.2	134
Box 7.4	Evaluation of soil vapour extraction project	141

Box 7.5	Monitoring and determination of effectiveness of soil flushing to remove cadmium	142
Box 7.6	Importance of monitoring soil vapour extraction systems for evidence of microbial activity	144
Box 7.7	Inspection of cover systems	147
Box A6.1	Possible drilling methods for installation of monitoring wells	173
Box A6.2	Examples of limitations of different drilling methods	174
Box A6.3	Surface geophysical methods used in the investigation of contaminated sites	178
Box A7.1	Key to Tables A7.2 to A7.6	191
Box A7.2	Gas chromatography	196
Box A7.3	Mass spectrometry	196
Box A7.4	Main reactions leading to instability of steel making slags	197
Box A7.5	Main reactions leading to instability of blastfurnace slags	198
Box A8.1	Examples of potential human health risks in contaminated land applications	201
Box A10.1	Application of groundwater modelling at a petrol station	212
Box A11.1	The Dutch ABC Values	214
Box A11.2	Application of Dutch Soil Protection Standards	215
Box A12.1	The US Environmental Protection Agency's Framework for Ecological Risk Assessment	222
Box A12.2	The concept of ecological risk assessment	222
Box A12.3	Basic concepts of ecology	225
Box A12.4	Causes of population decline due to contamination	227
Box A12.5	Effects of contaminants on community structure	228
Box A12.6	Biological factors governing the effects of contaminants on ecosystems	231
Box A12.7	Examples of assessment and measurement endpoints	234
Box A12.8	Measures to be used to compare areas under investigation with reference areas	235
Box A12.9	Bioassessment techniques	236
Box A12.10	Examples of measurements to be made on aquatic environments	237
Box A12.11	Temporal, spatial and other considerations in ecological risk characterisation	238
Box A12.12	Ecological epidemiology	238
Box A12.13	An example of an aquatic ecological risk assessment	239

List of Tables

Table 2.1	Examples of investigation objectives	10
Table 2.2	Phases of an investigation	12
Table 2.3	Terminology used to describe the phases of investigation	12
Table 2.4	Information required from the desk study	15
Table 2.5	Technical design aspects of the detailed investigation	20
Table 3.1	Specifications for ground investigation	29
Table 3.2	Example of measurement of 'total' metals by digestion with nitric/hydrochloric acids, followed by measurement by atomic absorption (AA) or inductively coupled plasma (ICP) spectrometry	36
Table 3.3	Health and safety measures that may be required for site investigations	37
Table 4.1	Design aspects of sampling	45
Table 4.2	Methods of exploration	47
Table 4.3	Sampling options for soils etc.	49
Table 4.4	Types of sample container	57

Table 4.5	Results of experimental studies on sampling soils for volatile organics	59
Table 4.6	Phasing groundwater investigations	62
Table 4.7	Types of groundwater monitoring installations	63
Table 4.8	Performance of sampling devices (after reference 25)	65
Table 4.9	Performance of sampling devices in relation to contaminant types	65
Table 4.10	Possible approaches to sampling surface waters	66
Table 4.11	Analytical design parameters	70
Table 5.1	Substances and materials which may be a hazard if present on a site	87
Table 5.2	Possible hazards, pathways and targets	87
Table 5.3	Characteristics of substances	88
Table 5.4	Classification of some organic chemicals	88
Table 5.5	Information sources on the impact of hazards to humans	90
Table 5.6	Matrix of hazard/pathway/target scenarios	92
Table 5.7	Possible screening factors for potentially contaminated sites	102
Table 6.1	Examples of non-dedicated guidelines and standards	111
Table 7.1	Compliance/performance testing for excavation based operations	131
Table 7.2	Monitoring and compliance testing of some in-situ treatment methods	140
Table 7.3	Possible permanent monitoring measures for containment systems	147
Table A2.1	Information sources likely to be used in the desk study	165
Table A2.2	Key information sources for hydrological data	166
Table A2.3	Sources of information on the relationship between site use and nature of contamination	166
Table A6.1	Possible construction materials	176
Table A7.1	Organisations producing standard analytical and test methods for soils and waters	189
Table A7.2	ISO standards relating to soil quality – terminology	190
Table A7.3	ISO standards for soils – sampling	191
Table A7.4	ISO methods for soil – chemical analysis	192
Table A7.5	ISO methods for soils – biological assessment	193
Table A7.6	ISO methods for soils – physical characteristics	194
Table A7.7	ISO methods for soils available as British Standards (January 1995)	194
Table A8.1	Risk of death from a range of common causes	202
Table A8.2	Risks associated with travel	203

1 Introduction

1.1 A RISK MANAGEMENT APPROACH

Risk management (see Box 1.1) provides an objective, iterative process for identifying, describing and evaluating the risks that may be associated with contaminated land and deciding the best way of controlling or reducing these risks, and implementing strategies to achieve acceptable levels of risk.

For the purposes of this Report the following definitions have been adopted:

Contamination is: the presence in the environment of an alien substance or agent, or energy, with a potential to cause harm.

Pollution is: the introduction by man into the environment of substances, agents or energy in sufficient quantity or concentration as to cause hazards to human health, harm to living resources and ecological systems, damage to structure or amenity, or interference with legitimate uses of the environment

---

Box 1.1 *A risk-based approach to contaminated land*

Risk management is defined as [1]:

'the process whereby decisions are made to accept a known or assessed risk and/or the implementation of actions to reduce the consequences or probabilities of occurrence'

It involves[2]:

1. Hazard identification.
2. Hazard assessment.
3. Risk estimation.
4. Risk evaluation.
5. Risk control.

Stages 1-4 inclusive are generally referred to as 'risk assessment'; stages 4-5 inclusive represent 'risk reduction'.

The principal advantages of risk management are that it:

- is structured
- is objective
- is comprehensive
- explicitly considers uncertainties
- provides a rational, transparent and defensible basis for discussion of a proposed course of action with (e.g. regulators, funders, insurers, the local community).

Site investigation produces the data needed to apply a risk management approach. That is, the data required to:

- enable contaminants or other hazards to be identified on a site-specific basis (hazard identification) and, in a wider context, to identify which sites out of a larger number are likely to pose the greatest risks (hazard ranking and/or preliminary risk assessment)

- assess the types and degree of hazard(s) which may come into contact along specified pathways with identified targets (hazard assessment)

- estimate (quantitatively or qualitatively) the likelihood that an adverse event will occur, and the magnitude of harm (risk estimation)

- judge the significance of the risks, taking into account any uncertainties in the assessment (risk evaluation)

- decide, where appropriate, the best way of reducing or controlling risks, and to implement and monitor the performance of any remedial measures put into place (risk control).

Site investigation is itself an iterative process; several phases of investigation may be required before sufficient data are available to fully characterise the hazards, pathways and targets of concern, and to estimate (even qualitatively) the risks involved. Extremely detailed information, possibly gathered over an extended period, will usually be required for a quantified estimate of risks, and even then significant uncertainties may remain. Supplementary investigations may be necessary to complete the risk assessment and/or support the choice of remedial methods. Investigation for compliance and performance purposes is essential to confirm that remedial action has achieved its objectives.

This Volume provides detailed information and guidance on:

- planning investigations so that they generate sufficient, good quality data for the effective management of risks (Section 2)

- implementing site investigations so that they are safe, effective and economic (Section 3)

- sampling and analysis to ensure that data obtained through site investigation are accurate and fully representative of site conditions (Section 4)

- assessing hazards and risks having regard to current guidelines and standards (Sections 5 and 6)

- investigations for compliance and performance purposes (Section 7).

This Volume refers to existing guidance on contaminated land[3], and extends existing geotechnical and hydrological guidance where particularly relevant to contaminated land. Some detailed technical guidance is provided in the Appendices, although this is not intended to be exhaustive.

It also focuses on the investigation of solid and liquid phase contaminants. Note that sites may also be contaminated with potentially hazardous gases, such as methane, carbon dioxide and hydrogen. For information on investigating such sites see references 4 to 6.

1.2 THE PURPOSE OF INVESTIGATION

Site investigation is essential to:

- determine the nature and extent of any contamination

- identify any hazards posed by the contamination

- identify and describe pathways and targets
- allow assessment of current and potential risks
- determine whether remedial action is needed
- inform decisions on the nature, extent and performance requirements of any remedial works.

A site may be investigated because of a proposal to redevelop, or the need to establish whether it poses environmental and/or human health risks, irrespective of current or future use. The investigation will vary in scope depending on the circumstances, but for a proposed redevelopment will usually address both contamination and engineering aspects.

Four key aspects should be addressed for the effective risk assessment of contaminated sites:

1. Contamination.
2. Hydrological (surface and groundwater) regime.
3. Geotechnical properties and geological setting.
4. Present and future targets and pathways.

On some sites more specialist studies such as ecological assessments may also be required (see Appendix 12).

The investigation must follow a formal strategy if it is to be carried out safely and effectively, with the maximum relevant information obtained and proper control over costs. A faulty plan can lead to the collection of invalid, inaccurate or irrelevant data, increased risks to health and safety, and potential environmental damage, as well as leading to much greater remedial costs. Figure 1.1 is a flow-chart showing the outline of a plan, including its various components and phasing, for the investigation of a contaminated site. This is discussed in more detail in Section 2.

The most important part of planning is setting the objectives for the investigation, since these are the guiding criteria for the strategy and its implementation. Objectives are largely site-specific, although typical objectives for different phases of the investigation are discussed in Section 2. Objectives should be set both for the investigation as a whole and for its component parts, e.g. on-site works, analysis and testing.

One of the principal differences between the investigation of potentially contaminated sites and those which are uncontaminated is the much greater consideration which must be given to health and safety issues. For this reason, and to maximise the quality of information obtained, no on-site (fieldwork) investigation should be done unless and until preliminary research (e.g. on the history of the site and existing conditions) has been carried out. Similarly, investigations for other purposes, such as construction, should not be carried out until at least some exploratory investigation of the contamination and hydrological characteristics of the site has been completed.

No matter how careful and comprehensive the site investigation is, there will always remain some uncertainty about the nature and extent of contamination, and the presence of other potentially hazardous conditions. In practice, site investigation must be regarded as a continuing process extending into the remediation stage. It is essential to make adequate contingency arrangements for any responsive action that might be required as the investigation progresses, to be observant during actual site works and to keep the remediation strategy under review at all times.

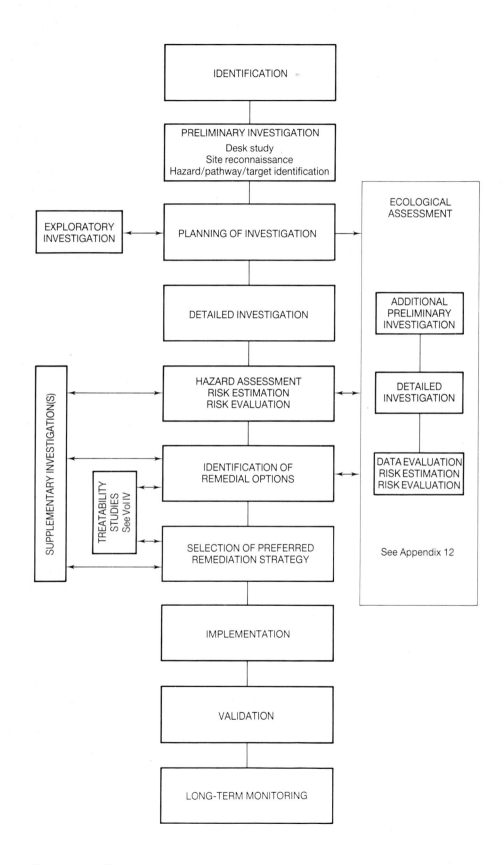

Figure 1.1 *Phased investigations for risk assessment and remedy selection*

Given the pivotal role of site investigation and assessment in ensuring both the effectiveness and safety of remedial action, it is vital that sufficient time and financial resources are committed to these elements of a remediation project. The first phase (desk study and site reconnaissance) should provide sufficient information on the scope of subsequent (e.g. on-site) phases of investigation for estimating the resources that will be needed to complete the work.

In 1991, the Institution of Civil Engineers (ICE) report *Inadequate Site Investigation*[7] highlighted the drawbacks associated with poorly-funded, inappropriately procured and insufficiently quality-controlled site investigation. Construction projects were found to have been delayed due to unforeseen ground conditions, expensive redesign was required in some schemes because of poor characterisation of the ground conditions prior to excavation, and uncertainty was found to lead to over-conservatism and unnecessarily expensive design measures. Although this survey addressed the general problems of defining ground conditions, its conclusions are highly pertinent to the investigation of contaminated sites. Subsequently the Site Investigation Steering Group published national guidelines on planning, procurement, specification and quality management to improve site investigation practice[8-11].

1.3 TYPES OF INVESTIGATION

This Volume differentiates between:

- investigations carried out for the purposes of assessing/controlling hazards and risks, and
- investigations carried out for construction purposes.

The term 'site investigation for hazard/risk assessment' is used to include all those activities that enable:

- potential hazards to be identified
- a hazard and/or risk assessment to be carried out
- a decision to be made on whether remedial action should be taken
- an informed choice to be made between different remedial strategies.

Whenever the terms 'site investigation' or 'investigation' appear in the text on their own, they refer to this type of investigation, unless the context makes clear that a more restrictive meaning is intended.

The term 'investigation for construction purposes' is used here to describe the types of investigation described in Box 1.2.

The approach described here relates principally to the investigation of 'typical' former industrial sites that have undergone clearance (decommissioning, decontamination, demolition following the procedures described in Volume II) and are required for redevelopment. However the same principles apply to the investigation of contaminated land in other contexts, such as infrastructure projects, on operating sites, for developed sites and 'problem' sites (see Volume X on Special Situations), or in the form of pre-purchase or pre-sale audits, or as part of a corporate environmental strategy.

> **Box 1.2** *Investigation for construction purposes*
>
> In a construction context, the term 'site investigation' refers to the process (including desk study) by which all relevant information concerning the site of a proposed civil engineering or building development, and its surrounding area, is gathered. The term 'ground investigation' defines[12] a narrower process, involving the acquisition of information on the ground conditions in, and around, a site.
>
> The CIRIA Site Investigation Manual[13] provides the following definitions:
>
> - **Site investigation is the** *'comprehensive investigation of a site, including past use and environmental constraints, to determine its suitability for development and to provide information which leads to safe, economical and practical designs of structures'.*
> - **Ground investigation is** *'exploratory investigation to determine the structure and characteristics of the ground influenced by a development. The collected information is used to establish or predict ground and groundwater behaviour during, and subsequent, to construction'.*

REFERENCES

1. ROYAL SOCIETY. *Risk : Analysis perception and management.* ROYAL SOCIETY (London), 1992

2. PETTS, J, Risk assessment for contaminated sites. In : *Proceedings of a Conference on Site Investigations for Contaminated Sites.* IBC Technical Services (London), 1993, Paper No 1.

3. BRITISH STANDARDS INSTITUTION. *Draft for development Code of Practice for the identification of contaminated land and its investigation.* DD 175, BSI (London), 1988

4. CROWHURST, D. and MANCHESTER, S.J. *The measurement of methane and other gases from the ground.* Report 131, CIRIA (London), 1993

5. HOOKER, P.J. and BANNON, M.P. *Methane: Its occurrence and hazards in construction,* Report 130, CIRIA (London), 1993

6. RAYBOULD, J.G., ROWAN, S.P. and BARRY, D.L. *Methane investigation strategies* FR/CP/14 CIRIA (London), 1993

7. INSTITUTION OF CIVIL ENGINEERS. *Inadequate site investigation.* Thomas Telford (London), 1991

8. SITE INVESTIGATION STEERING GROUP. *Site investigation in construction 1: Without site investigation ground is a hazard.* Thomas Telford (London), 1993

9. SITE INVESTIGATION STEERING GROUP. *Site investigation in construction 2: Planning, procurement and quality management.* Thomas Telford (London), 1993

10. SITE INVESTIGATION STEERING GROUP. *Site investigation in construction 3: Specifications for ground investigation.* Thomas Telford (London) 1993

11. SITE INVESTIGATION STEERING GROUP. *Site investigation in construction 4: Guidelines for the safe investigation by drilling of landfills and contaminated land.* Thomas Telford (London) 1993

12. BRITISH STANDARDS INSTITUTION. *Code of Practice for Site Investigations*. BS 5930:1981. BSI (London), 1981

13. WELTMAN, A. J. AND HEAD, J. M. *Site investigation manual*. Special Publication 25. CIRIA (London), 1983

2 Planning the investigation

2.1 SCOPE

When planning an investigation both technical and operational aspects should be taken into account. Particular attention should be given to phasing (see Section 2.3), and the integration of different types of investigation activity. All plans should make provision for:

Technical aspects

- the types of data to be collected (e.g. on contamination, hydrology, geology)
- how the data are to be used (e.g. to identify and characterise hazards, targets, pathways, and to assess risks)
- sampling and testing requirements (e.g. techniques, strategies, management, specification etc.)
- phasing of the investigation
- integration of different investigation activities.

Operational aspects

- health and safety protection for site personnel and the general public
- the use of appropriate procurement methods and contractual conditions
- minimising environmental impacts and residual liability
- legislative and regulatory requirements
- communication with interested parties.

Detailed guidance on site investigation is available in various published standards and documents, (see Appendix 1). The investigation of contamination should conform with the recommendations of the British Standard Draft for Development, DD175:1988, *Code of Practice for the identification of potentially contaminated land and for its investigation*[1]. This will be supplemented by a variety of ISO (International Organisation for Standardisation) standards on soil sampling which are being prepared. The geotechnical aspects of the investigation should conform with BS 5930[2] (currently under revision — January 1995). Guidance should also be sought from the documents on site investigation in construction[3] (published in 1993). Further guidance on the investigation of derelict land can be found in CIRIA Special Publication 78 *Building on Derelict Land*[4]. Guidance on the investigation of the water environment is available in a range of British/International Standards (see Appendix 1).

2.2 SETTING OBJECTIVES

Objectives must be decided on the basis of the available information, the implications of anticipated conditions for the proposed or actual use of the site, and the potential hazards and risks which could arise from the site conditions. Allowance must be made within the investigation plan for regular review and modification of the objectives where necessary to take account of the findings of each phase of the work.

The main objective defines the purpose of the work; subsidiary objectives relate to specific aspects. The main objective might be:

'To provide sufficient information on potential hazards, pathways, targets and other site characteristics (e.g. engineering constraints) to permit an assessment of risks and allow decisions to be made on the need for, and nature of, any remedial work including any immediate action to protect public health or the environment.'

Satisfying the main objective involves:

- determining the nature and extent of any contamination of soils and groundwater on the site
- determining the nature and extent of any contamination migrating off the site into neighbouring soils and groundwater
- determining the nature and extent of any contamination migrating into the site
- determining the nature and engineering implications of other hazards and features on the site e.g. expansive slags, combustibility, deep foundations, storage tanks
- identifying, characterising and assessing potential targets and likely pathways
- providing sufficient information (including a reference level to judge effectiveness) to identify and evaluate alternative remedial strategies
- determining the need for, and scope of, both short- and long-term monitoring and maintenance
- formulating safe site working practices and ensuring effective protection of the environment during remedial works
- identifying and planning for immediate human health and environmental protection and contingencies for any emergency action.

It also requires specific objectives to be set for each phase of the investigation, (see Section 2.3) and for each of four principal aspects:

1. Contamination.
2. Hydrology (i.e. the whole of the water environment).
3. Geotechnics and geology.
4. Target and pathway identification/characterisation.

Table 2.1 lists typical objectives for each of these aspects.

Hydrological data are important because they describe surface water bodies and conditions above (the *vadose* zone) and below the water table (the saturated zone). They can also be used to predict future concentrations and movement of the contaminants. Since limited site investigation provides only a 'snap-shot' of a dynamic situation, long-term monitoring of the contamination profile and surface/groundwater conditions will be needed for a full understanding of the hydrological regime and its likely relevance to, and influence on, any remediation strategy.

Table 2.1 *Examples of investigation objectives*

Aspect	Objectives/information needs
Contamination	To determine the: • nature, extent, source and distribution of contaminants (on and off-site) in a range of media — soil/fill/wastes, ground/surface water, air, biota, containers (drums etc.) • form of contamination or contaminated media — gaseous, liquid, semi-solid, solid • ground temperatures • level of microbial activity • health of ecosystems (soil, water, land area)
Water environment	To determine where appropriate the: • groundwater levels/pressures and their variation with time • direction and volume of flow of ground and surface water • abstraction and recharge activities having an influence on the site • chemical and mineralogical quality of ground and surface water • background chemical composition of surface and groundwater in the area • geological strata composition and structure • primary and secondary permeability/porosity • propensity of site to flood • rainfall and evaporation characteristics • tidal fluctuations
Geotechnics	To determine, where appropriate, the: • physical characteristics of the ground e.g. presence of in-ground obstacles, services etc. • physical characteristics of contaminated matrices e.g. mineralogy, moisture content, permeability, chemical composition, particle size distribution • geotechnical characteristics e.g. strength, compressibility, stability of slopes, existing structures, potential for subsidence etc. • presence of old mine workings
Actual and potential targets and pathways	• potential exposure pathways identified from detailed analysis of all above information • existing or proposed use of site and surrounding land • potential human targets including site workers (investigation/remediation/construction/maintenance), occupants, users, neighbours, trespassers • proximity to sensitive ecosystems • proximity of water bodies • proximity to economically valuable natural resources (e.g. mineral deposits)

> **Box 2.1** *Geotechnical parameters to be addressed by the investigation*
>
> The investigation should gather all data needed for assessing:
>
> - the general suitability of the site and neighbourhood for the proposed remedial works
> - the need for and design requirements of any foundations (e.g. for treatment plant), earthworks and temporary works associated with the remedial strategy, taking into account the effect of any previous uses of the site
> - any factors arising from the soil or groundwater conditions that might constrain the construction or implementation of remedial works including temporary works, excavation, trafficability and drainage
> - the quantity, quality and ease of extraction of construction materials (e.g. concrete foundations) suitable for inclusion in the works
> - changes in the stability, drainage or other geotechnical aspects of the site and the surrounding ground and buildings which might be initiated by the remedial works.
>
> In addition, it may be necessary to assess the stability of existing structures, or to evaluate potential failure or instability.

An understanding of the geological/geotechnical characteristics of the ground within and around the site is essential for prediction of contaminant behaviour, and hence identification of susceptible targets and potential risks. For example, both the solubility of a contaminant and the physical nature of the ground affect contaminant mobility and hence availability to potential targets. Knowledge of the physical characteristics of contaminated matrices is also important in the selection and successful implementation of remedial methods. Properties such as mineralogy, moisture content, permeability etc. are as important as chemical composition in determining the applicability or otherwise of the various methods. The geotechnical properties of the site also have practical implications for the implementation of remedial works, e.g. the need to provide physical support during excavation, material preparation/handling facilities (crushers, size classifiers) prior to treatment, or the presence of in-ground obstacles and other features. Box 2.1 lists some important geotechnical parameters to be considered when planning investigations.

2.3 PHASING OF THE INVESTIGATION

Investigations for hazard/risk assessment purposes typically involve the phases listed in Table 2.2 and described in Sections 2.4 to 2.6 (note that the terminology used to described the phases of an investigation may vary — see Table 2.3).

Phasing is particularly important in the investigation of contaminated land, because of the need to prevent harm to the public, those working on the site and the environment.

The principal phases of investigation are:

Phase 1 Preliminary investigation — desk study and site reconnaissance

Phase 2 Exploratory investigation

Phase 3 Detailed investigation and assessment.

Intermediate phases may be required for specific purposes, e.g. to aid hazard, target, or pathway identification/characterisation, or to satisfy different technical objectives (e.g. contamination, hydrology or geotechnical characterisation).

A further two phases of work may be required before all investigative work on a contaminated site can be considered complete (see Table 2.2).

Table 2.2 *Phases of an investigation*

Phase	Objectives
Preliminary investigation (comprises desk study and site reconnaissance)	Provides background information on present and past uses, and the nature of any hazards
	Provides background information on the geology and hydrology of the site
	Provides an initial overview of the nature and extent of contamination
	Identifies targets or features of immediate concern
	Informs decisions about the need for specialist ecological assessment
	Provides data to inform the design of exploratory and detailed investigations and an early indication of likely remedial needs
	Provides information relevant to worker health and safety and protection of the environment during on-site investigations
	Can provide data for hazard ranking and preliminary risk assessment (see Section 5)
Exploratory Investigation	Intended to confirm initial hypotheses about contamination and site characteristics, and to provide additional information to aid design of detailed investigation(s), including health and safety etc. aspects
Main (detailed) Investigation	To characterise fully the extent of contamination, the hydrology and geology
	Provides data required to inform the hazard and/or risk assessment and the selection of remedial methods
Supplementary Investigation	Where required, addresses or clarifies particular technical matters (e.g. to confirm the applicability and feasibility of one or more potential remedial options) or to collect information relevant to the application a of selected remedial option
Investigations for compliance and performance	Seek to confirm proper implementation and effectiveness of remedial measures

Table 2.3 *Terminology used to describe the phases of investigation*

Activity	This report	BS DD 175 1988[1]	ISO standards	BS 5930 1981[2]	Environmental Audits	Other terms
Collection of historic, published and anecdotal information	Preliminary investigation:	Preliminary investigation:	Phase 1		Phase 1	Site characterisation
	Desk study	Desk Study	Desk Study	Desk Study		
Visit to site	Site reconnaissance	Site reconnaissance	Site reconnaissance	Site reconnaissance		Walkover survey, preliminary inspection
Initial limited on-site investigation	Exploratory investigation	Not specifically identified	Phase 2	Not specifically identified	Phase 2	
Full comprehensive on-site investigation	Detailed investigation	On-site investigation	Phase 3	Ground investigation, also detailed examination and special studies	Phase 3	
Further, more targeted investigation	Supplementary investigation	Not identified	Phase 4	Not specifically identified but could be special studies	Phase 4 and subsequent	
Confirmation that remedial measures have been carried out as specified and are effective	Investigation for compliance and performance	Not identified	Not identified	Not identified	Not identified	

Supplementary investigations may be required to refine the choice of remedial methods in two main ways:

1. by gathering specific information on contamination and other site characteristics to help determine on the applicability and feasibility of one or more remedial options.

2. by gathering information vital to the successful implementation of a selected option, e.g. to confirm assumptions about soil characteristics, volumes and amounts of contaminants to be treated, or the geology along the line of an in-ground barrier.

Investigation for compliance and performance are required to confirm that remedial action has been successful.

The design and implementation of these subsequent phases should conform to the general principles and procedures outlines for the detailed investigation in Section 2.6. Objectives should be defined, methods clearly specified and findings thoroughly documented.

The detailed requirements of investigations for compliance and performance are described in Section 7.

For guidance on the conduct of investigations as part of ecological assessments see Appendix 12. Specialist advice should always be sought when such investigations are required.

Post-closure surveys will be needed when preparing plans for the decommissioning, decontamination and demolition of plant and buildings, as discussed in Volume II. It may be necessary to determine:

- the nature of surface deposits
- the contents of tanks, pipes, drums etc.
- the nature and extent of contamination of the building fabric.

Where investigation of the site itself, or surface deposits, it required, the procedures described in this Volume will be applicable. More specialist aspects (e.g. the investigation of the fabric of buildings for contamination) are dealt with in Section 3 of Volume II.

2.4 PRELIMINARY INVESTIGATION

2.4.1 Scope

The preliminary investigation consists of a formal desk study and a site reconnaissance. The aim is to obtain initial information on the actual and probable nature and location of contamination and other hazards, present and future potential targets and, where appropriate, relevant geological, geotechnical and hydrological (including hydrogeological) features. It may also be the basis for 'hazard assessment', 'hazard ranking' or 'preliminary risk assessment' (see Section 5).

The preliminary investigation provides the information needed to decide site-specific investigation objectives and procedures for the subsequent on-site investigation. Determining the health and safety and environmental protection requirements for on-site work is an important task of preliminary investigation.

Preliminary investigation should also reduce the risks of:

- an investigation design which requires the comprehensive measurement of contaminants and other hazards which, in reality, are unlikely to be present or relevant to the objectives of the investigation
- an inadequate investigation design which fails to provide the data needed either to assess the hazards and risks or to select appropriate remedial measures where necessary.

Preliminary investigation is sometimes followed shortly afterwards by an exploratory investigation to provide limited data on the actual nature of the contamination at the site, and in particular to test hypotheses developed on the basis of the preliminary investigation (see Section 2.5).

The initial appraisal may generate sufficient information to assess the suitability of the proposed development, any major practical constraints and the probable order of costs. It may also indicate any sensible modifications, either in design or layout, to accommodate conditions on the site, for example relocation of structures away from particular problem areas.

The importance and value of the preliminary investigation should not be underestimated. Costly mistakes and inappropriate action can be avoided and risk significantly reduced through an early, albeit not comprehensive, understanding of the extent and nature of contamination on the site. Box 2.2 summarises the typical output of the preliminary investigation.

Box 2.2 *Output from the preliminary investigation*

The output from the preliminary investigation should be a report that:

- collates and summarises all relevant information (e.g. on site history, geology, hydrology, plans for development of the site and neighbouring areas)
- describes the methodology employed
- provides clear statements of the conclusions (hypotheses) about the anticipated contamination of the site (nature, location etc.), the hydrological regime and the geological setting
- identifies potential hazards and targets currently at risk or that would be at risk under a range of defined scenarios (e.g. no action, development for a specified purpose)
- identifies immediate hazards to public health or the environment and details any emergency action that should be taken (e.g. investigation, fencing, removal)
- outlines the overall strategy and objectives for subsequent investigation, including recommendations as to whether an exploratory investigation is required (prior to design and implementation of a main/detailed investigation) to confirm that the hypotheses/conclusions drawn are reasonable.

The report should be made available to all those concerned with subsequent phases of investigation. It should accompany any further reports on the site.

2.4.2 Desk study

The desk study involves the collection and assimilation of information on:

- the history of the site, including the nature of any industrial processes and other activities that are likely to have contaminated, damaged or otherwise degraded the site
- the geological and hydrological setting

- present and future potential targets (e.g. children trespassing on the site, a nearby water abstraction point).

Where significant time has elapsed between the active or contaminative use of the site and its subsequent investigation, or where access to information regarding a site is not readily available, the desk study will normally rely on the collation of historical documentary and other information on the use of the site. For operating sites, a similar exercise should be undertaken, but in this case based on the detailed knowledge of operational characteristics available to site personnel. Of particular value, and increasingly likely to be available, are the results of environmental audits carried out while the site was in active use.

Examples of information sources are given in Appendix 2. BS DD 175:1988[1] and BS 5930[2] provide guidance on desk study procedures. Reference can also be made to Building Research Establishment publications (see Appendix 1). Information on the relationship between land use and contamination is available from a number of sources (see Appendix 2). Examples of the type of information required from the desk study are given in Table 2.4.

Table 2.4 *Information required from the desk study*

Item of information	Examples
Site layout (as built)	Plant components, building structures, drainage systems, process areas, storage areas, energy supply plant, effluent treatment plant, gas treatment plant, waste disposal plant and areas, maintenance facilities, laboratory facilities, site services
Design/construction modifications.	Site layout, process train, materials
Nature/quantities of materials handled on the site	Feedstocks, intermediates, products, wastes, reagents, maintenance materials
Nature of surrounding land use	Residential, hospitals, schools, nurseries, commercial/industrial, agriculture/horticulture, surface/groundwater resources, general ecology, ecologically valuable habitats
Physical features	Present and past topography, propensity for flooding
Previous history	Industrial use, incidence of major accidents (fires, spillages, leaks)
Nature of building fabric and structural condition of plant/buildings	See Volume II
Geology/hydrology	Superficial and drift geology, presence/status of surface and groundwater bodies

It is important to collect information on hydrological aspects as part of the desk study. Since on-site investigation activities may themselves cause inadvertent contamination of an aquifer or water course, it is essential to have a good understanding of the hydrological regime as a whole, irrespective of whether current water quality is an issue at the site.

Hydrological and geological data gathering should be coordinated because of the close relationship between the two aspects. Information sources of particular relevance to hydrology are listed in Appendix 2. Site-specific data on water quality are rarely available and reference to past and present uses of the site and adjoining land is essential for an early assessment of likely water quality beneath and around the site. Note the possibility that contaminants may have migrated from sources at some distance from the area of investigation.

2.4.3 Site reconnaissance

The site reconnaissance involves a visit to the site and neighbouring land. It should not be undertaken until the desk study information has been collated and assessed, so that investigation personnel are not put at unnecessary risk. Prior to carrying out the visit suitable maps and plans should be prepared, the team briefed on potential hazards, and a strategy for the reconnaissance devised. Visual evidence of contamination and other hazards can confirm or otherwise supplement information obtained from the desk study. Identification of susceptible targets and immediate hazards is an important objective of this stage. Existing boreholes etc. should be located. The survey should, wherever possible, extend beyond the boundaries of the site to look for evidence of contaminant (or gas) migration.

Where sites are obviously severely contaminated (e.g. leaking drums etc. lying around) or there are obvious physical hazards, an initial survey from outside the site boundary using binoculars, television cameras etc. may be advisable to help build up an appreciation of site conditions.

Site reconnaissance should normally be made by a small team which includes, when possible, someone familiar with the site (e.g. a former employee) and someone who can appraise the structural safety of any remaining structures (see Volume II and reference 5). The site reconnaissance typically includes the activities listed in Box 2.3 (see also Section 3 of Volume II).

A limited amount of sampling may be undertaken including from surface deposits, existing boreholes, surface water courses (and springs). This should not be confused with formal exploratory investigation which can only be designed and carried out once the results of the preliminary investigation have been evaluated and conclusions drawn about the nature and distribution of contaminants, and the geology and hydrology of the site (see below).

All on-site observations must be accurately recorded, for example by accurate location of observation points on a map with features described in a notebook and/or on tape for later transcription. Where appropriate, markers can be left on the site to aid relocation of particular features.

2.4.4 Topographical survey

Most investigations will require a topographical survey to be carried out. Provided there are no health and safety constraints, this should be done at an early stage as part of, or on completion of, the preliminary investigation and prior to any on-site works, so that individual site features and sampling points can be accurately located.

2.4.5 Development of hypotheses for subsequent investigation phases

The data obtained from the preliminary investigation should be used to develop hypotheses regarding the presence and distribution of contaminants at the site. The on-site investigation can then be designed to test the hypotheses, thereby increasing the chances of confirming the presence of contamination and permitting economy of effort. The hypotheses developed should embrace all aspects of the investigation but in particular should address:

- whether it is appropriate to 'zone' the site for the purpose of subsequent investigation, e.g. on the basis of past land use or topography[6]

- whether it is possible to draw conclusions about the probable distribution of contamination, either within the site as a whole or within defined zones.

> **Box 2.3** *Site reconnaissance activities*
>
> - Noting any obvious immediate hazards to public health or safety (including to trespassers) or the environment
> - Noting the condition of any fences etc. and any other factors affecting the security of the site
> - Noting any areas of discoloured soil, polluted water, distressed vegetation or significant odours
> - Confirming as far as possible the location of buildings, roads, fences etc. and any deviations from those shown on the available plans
> - Noting the location and condition of any remaining buildings, structures, tanks etc.
> - Recording the presence, location and condition of any surface deposits and made ground, and any signs of settlement, subsidence or disturbed ground
> - Determining the depth of any standing water, and the direction and rate of flow of water in any rivers, streams or canals
> - Noting any evidence of gas production or underground combustion
> - Confirming the location of sewers etc.
> - Noting the location of services including telephones, water and electricity, that might be required or damaged during site investigation
> - Noting/confirming the uses of neighbouring land and, in particular, any activities that may have led to contamination of the site under investigation
> - Noting/confirming the position of any outfalls to surface water and the nature/condition of any discharges
> - Looking for any evidence of seepages through river or canal banks etc.
> - Locating and noting the condition of any boreholes etc. remaining from previous investigations
> - Noting where access can be gained to the site for site investigation equipment, waste disposal vehicles etc.
> - Locating areas suitable for use during investigation for depots, offices, laboratories sample storage, etc.
> - Use of suitable portable instrumentation to determine the presence, and possible concentrations of hazardous gases
> - Limited sampling of surface deposits, surface waters etc.
> - Making a photographic record of general site conditions and layout, and of any individual important features.

The development of formal hypotheses is addressed in NVN 5740[7]. It stresses the need to have a hypothesis regarding the nature and distribution of contamination, and other important factors such as the heterogeneity of strata and the hydrological regime, before deciding the strategy for an exploratory investigation. A similar approach is likely to be adopted in the comparable ISO guidance[8].

Typical hypotheses which may be applied to a whole site or to parts of a site (i.e. different hypotheses for different parts of the same site) are detailed in Appendix 3, but could include the premise that part or all of the site is uncontaminated. Alternatively, the hypotheses might relate to a certain distribution of contamination within the site. NVN 5740 makes detailed recommendations on investigation strategies for exploratory investigations (including the number of sampling points, number of samples to be tested etc.) for some of the more common hypotheses. Sampling strategies are discussed in Section 4.

Hypotheses about the spatial distribution of contamination should take into account:

- the nature of the source and the manner in which the contamination has entered the soil, e.g. diffuse (aerial deposition, deposit of sewage sludge etc.), or spot (e.g. leaking tank).

- where in the soil or the groundwater the contamination is likely to be found, taking into account the expected migration processes in both vertical and horizontal directions. These will depend on:

    - the nature of the contaminants (e.g. solubility in water, interaction with clays and organic matter in the soil)

    - soil stratification (nature and permeability of soil/fill, thicknesses of strata)

    - soil/fill characteristics such as pH, redox potential (Eh), organic matter content, and cation exchange capacity

    - the depth to groundwater and conclusions drawn about probable direction of flow, and temporal variations of both of these

    - how long the contamination has been in the ground.

2.5 EXPLORATORY INVESTIGATION

Exploratory investigations are carried out principally to test the truth of hypotheses (i.e. concerning the probable presence, nature and extent of contamination; the physical characteristics of the site; and/or the water environment) derived from the preliminary investigation, and to amend them if appropriate. By their nature exploratory investigations are limited in scope, and will not usually be sufficient to prove the absence of contamination. Locations for boreholes and sampling points etc. and the analytical strategy, should be designed to minimise resource and time requirements and to maximise the chances of confirming the hypotheses rather than to build up an overall picture of site conditions. The results of such an investigation may confirm expectations but a failure to confirm expectations will not usually mean that the site is uncontaminated: rather that the hypotheses are wrong. Only a more detailed investigation can resolve the resulting questions.

An exploratory investigation might be used to help design the detailed investigation, or in some cases, might be considered as sufficient for a pre-purchase or pre-sale audit, depending on the level of risk of incurring environmental liabilities with which the parties to the transaction are comfortable.

To limit the overall costs, whenever possible the exploratory investigation should be designed to contribute towards the site investigation as a whole (i.e. form a first phase of the detailed on-site investigation) rather than as a completely separate entity. Reference should be made to Tables 2.4 and 2.5 for information on the design and planning elements of on-site investigations.

Although for contaminated site investigations 'spot' samples of soil are usually taken for analysis, it may be appropriate on occasion to take composite samples or to prepare mixed samples (see Box 2.4) representative of a small area of the site during an exploratory investigation (e.g. on former farmland or elsewhere when there is an expectation that the contamination may be fairly evenly spread). If this is done, any guidelines used to judge whether contamination is present or of concern should be revised downwards to allow for 'dilution' effects.

> **Box 2.4** *Soil sample types*
>
> **Spot sample:**
>
> Sample from a discrete location and depth
>
> **Composite sample:**
>
> Sample representative of a defined area prepared in the field by combining a number of sub-samples. Commonly used to evaluate the quality of agricultural land.
>
> **Mixed sample:**
>
> Sample prepared in the laboratory by mixing sub-samples taken from a number of spot samples (individual spot samples are retained for subsequent analysis if required). Samples combined in this way should come only from the same plane or strata type as appropriate.

2.6 DETAILED INVESTIGATION

2.6.1 Scope

This phase of the investigation provides detailed information on the nature and degree of contamination, the physical (including water) environment, and presence of gas and other hazards. The design of the detailed investigation should be based on the information obtained from the preliminary, and any exploratory, investigations in accordance with the objectives specified for this phase of the work. Sufficient detail is needed for:

- assessing health and environmental hazards and risks
- evaluating financial and technical options for the subsequent development if one is planned
- selecting and planning any remedial work
- designing the works
- ensuring safe working for personnel on-site
- ensuring health and safety of the public
- assessing requirements for both short- and long-term monitoring.

The actual scope of the detailed investigation will be highly site-specific and it is not possible to generalise here on what level of detail and hence what size/scope of investigation will be required. Where possible, guidance on sampling and testing strategies and protocols are given in Section 4 and Appendix 6. It is important to bear in mind that different contaminants may require different detection procedures because they:

- have different physical properties (e.g. viscosity, solubilities, specific densities, volatilities)
- interact differently with different soil components
- are affected differently by factors such as pH and redox potential
- may exhibit markedly different distributions.

Table 2.5 lists the principal aspects to be considered during the design of the investigation.

Table 2.5 *Technical design aspects of the detailed investigation*

Aspect	Variable
Exploratory excavations	Number Type (borehole, trial pit, trench, groundwater monitoring installation) Dimension Depth Location Method of installation
Sample collection	Media type (liquids, solids, gases, other) Number of samples Size and type of samples Location Depth Storage and handling arrangements
In-situ testing	Media type Determinand/property Number Location Duration of testing
Laboratory testing	Media type Sample preparation Determinand/property Analytical methods Test methods
Health and safety requirements	Procedures Protective equipment
Environmental protection	Air quality Ground and water quality monitoring Waste disposal Operating procedures

2.6.2 Planning

In addition to the technical aspects of design the following operational issues should be covered:

- access to working sites and adjacent land
- location and nature of in-situ obstructions
- location and status of services
- suitable areas or zones for offices, laboratory, sample storage, decontamination unit etc.
- ownership of all working areas
- water and power supplies
- communications and emergency services
- accident/emergency action plans
- action to protect the environment, e.g. dust suppression, wheel washing
- waste disposal arrangements for water, imported chemicals and for excavated waste and samples.

For specific sites and under particular circumstances there could be a number of further considerations e.g. availability of skilled workforce (see Section 3).

2.7 INTEGRATION OF INVESTIGATION PROCEDURES

2.7.1 Integration of different technical aspects

There are a number of benefits to be gained (see Box 2.5) from an investigation which addresses contamination, geotechnical and water environment aspects on an integrated basis. The need to take health and safety and environmental protection into account in the design and execution of geotechnical investigations, and the opportunities to consider all the data together, are particularly relevant. Depending on the circumstances of the site and timing of the investigations, integration can be achieved either by close liaison between specialists, or by employing personnel experienced in both areas.

Box 2.5 *Benefits of integrated investigations for risk assessment*

- project management may be simplified
- common use of equipment and procedures
- use of the same exploratory excavation or monitoring points for descriptive and sampling purposes (subject to any limitations on sample handling, storage, preparation etc.)
- use of geotechnical boreholes for subsequent installation of gas and/or water quality monitoring points
- ability to take account of the implications for health and safety arising from the presence of contamination for both geotechnical and hydrological studies
- allows an early understanding of the effect of geotechnical and hydrological factors on contaminant behaviour
- permits allowance to be made for geotechnical constraints on the treatment of contaminants
- provides for a combined consideration of the data obtained for each aspect on completion and as the investigation progresses

---

Much of the equipment, and many of the procedures used are similar for all three types of study. Data requirements and field observations are also interrelated. In some situations, these similarities can be exploited so that the same exploratory hole can be used for a number of data collection and sampling purposes. For example, it may be feasible to use a borehole drilled for groundwater monitoring purposes to provide information and samples of value to both contamination, and geotechnical studies; boreholes sunk to prove the depth of suitable founding strata are likely to provide information on the hydrology of the site. All additional excavations made on the site, such as pits to determine the location of live services, should be used to provide extra information, if only of a visual nature, on the contamination, geotechnical and hydrological profile of the site.

When considering an integrated approach to the investigation of a contaminated site, attention should be given to the technical objectives of each aspect, quality assurance procedures and project management. Priorities should be assigned where necessary. This is particularly important in relation to: the type and positioning of exploratory excavations; the sample collection and handling techniques used; the phasing of the investigation; and health and safety and environmental protection provision.

2.7.2 Integration between investigations for risk assessment and construction purposes

On a development or infrastructure site it may be possible to integrate investigations carried out for the purposes of assessing hazards and risks with investigations required for construction purposes. This applies particularly to geotechnical aspects. **Note, however, that the purposes of the two investigations are different and the objectives of each should not be compromised as a result of a combined approach.**

On particularly 'sensitive' sites, for example where contamination is prevented from reaching an underlying aquifer by a clay layer, the fewer excavations and boreholes that are placed the better, and there may be benefits other than just economy in combining the two types of investigation. It should be borne in mind that sensible decisions about how to develop such a site should really await the results of the risk assessment.

Examples of when integration is and is not acceptable are:

Acceptable

- Taking the opportunity for installation of trial pits or boreholes during an investigation for construction purposes to obtain additional soil and/or groundwater samples to provide supplementary information on contamination or the hydrogeology to better inform the risk assessment and the selection and design of remedial methods.

Not acceptable

- The relocation of a borehole critical to the understanding of contamination conditions (e.g. to intercept a contaminated groundwater plume) to provide geotechnical data on a part of the site designated for construction (an additional borehole will be needed).

- The use of large (up to 25 kg) disturbed samples originally collected for geotechnical testing purposes (and collected/stored/handled accordingly) for subsequent detailed organic analysis.

The exact details of how the two investigations can be combined in any individual project must depend on the specific requirements and objectives of each investigation. There are generalised areas where joint study is normally beneficial, including:

- the preliminary investigation including both desk study and the site reconnaissance
- the use of trial pits and excavations to inspect and sample strata
- the use of boreholes to provide samples for both contamination and geotechnical testing
- any analysis required for construction purposes being done as part of the overall analytical programme.

Combined preliminary investigation:

This is an essential part of both investigations and is most effectively carried out as one operation. The findings can then be used to:

- design an integrated (contamination, geotechnical, hydrological) investigation
- design the investigation for construction purposes
- establish health and safety requirements for each
- establish the degree to which the two investigations can be carried out jointly

- make preliminary assessments of the significance of contamination, hydrological and geotechnical constraints on the development of the site
- identify any constraints (e.g. depth of boreholes) imposed by the presence of contamination on the construction investigations.

Careful consideration of the objectives, strategy and procedures for each type of investigation, coupled with close liaison and understanding between the specialists involved should make possible a rational and economic approach to the investigation as a whole.

Combined exploratory holes

There should be few difficulties in using open excavations for both geotechnical and contamination investigations. However, using the same borehole for sampling for contamination and geotechnical purposes, and perhaps subsequently for groundwater and gas monitoring requires caution. Sampling methods are different (e.g. small (say 1 kg) 'spot' samples representing a discrete depth for contamination versus larger samples representing a depth range for geotechnical purposes) and storage/handling procedures vary depending on the type of analysis being conducted. Great care needs to be exercised if this is done, to ensure that neither of the investigations is compromised.

2.8 REPORTING

Comprehensive and accurate reporting of site investigation work is fundamentally important. Reports must describe the findings of the investigation and document the decisions made and actions taken at all stages of the investigation. Site investigation reports may be required for different purposes and to serve the needs of a number of different parties and different purposes. Those preparing site investigation reports should be mindful that they may be viewed by many different people, only a few of whom may have detailed technical knowledge. They may be passed to various regulatory agencies, used in evidence at a public enquiry or otherwise come under close scrutiny. It is vital, therefore, that a high standard of presentation is achieved and that the results and their assessment are presented with clarity and precision. Any constraints imposed on the investigation should be clearly identified.

Box 2.6 provides a brief summary of the typical contents of a report on the investigation of a potentially contaminated site. Box 2.7 contains some comments on presentation.

2.9 ECOLOGICAL ASSESSMENT

An ecological assessment is a specialist investigation of the ecology of a site and surrounding areas. It enables actual effects associated with contamination to be identified, potential future effects to be estimated and the need for any corrective action to be assessed. Ecological assessments are discussed in Appendix 12. Specialist advice should always be sought when such an assessment may be required.

> **Box 2.6** *Typical content of a site investigation report*
>
> 1. Summary. A description of the work carried out, stating the aims and main findings together with their implications and a brief account of the conclusions and recommendations.
> 2. Table of contents. This should include lists of tables, figures and appendices.
>
> *Factual*
>
> 3. Introduction. Details of the location and history of the site including a brief description of its current state and condition, the objectives of the investigation, and conclusions (hypotheses) from the preliminary report (see Section 2.4) used to guide the design and conduct of the investigation. Any important additions or changes to the picture of the site drawn from the preliminary study should be noted. The report should make clear the basis of the investigation strategy and whether there were any constraints imposed for practical reasons or because of a need to limit expenditure. It should be made clear why these constraints have been applied.
> 4. General matters. The report should describe general site conditions (e.g. weather), while the work was in progress, health and safety procedures and any other matters affecting the conduct or outcome of the investigation. Photographs and drawings showing the general layout of the site and conditions should be included (remember that the report may be read by people who have not seen the site).
> 5. Sampling. An explanation of the sampling strategy adopted, including plans showing the positions of the sampling points and their relation to the site history should be given. Any forced deviations from original plans should be noted together with the reasons. A description of materials encountered in trial pits or boreholes etc. should be provided (e.g. through inclusion of trial pit and borehole strata and sample logs). The methods employed to obtain, contain, stabilise, and handle samples etc. should be recorded. Photographs should be included where appropriate.
> 6. On-site testing. A full account should be provided of any on-site testing or measurements made including the methods employed and the results obtained.
> 7. Analysis and off-site testing. The sample preparation and sub-sampling procedures should be described. It is not necessary to give the analytical or test methods in full, unless they are novel, but a general indication of the method should be given including reference to the relevant British or other standards where appropriate. Information on the performance characteristics of the analytical or test method should be given where possible. All analytical and test results should be included in the report, although it may be more practical to include the detailed results in appendices and to present the results through summary tables and suitable visual presentations.
>
> *Interpretative*
>
> 8. Narrative description and discussion of results.
> 9. Data evaluation. Assessment of data in terms of adequacy of type, quantity and quality
> 10. Hazard Assessment
> 11. Risk estimation
> 12. Risk evaluation
> 13. Conclusions and recommendations

2.10 POST-CLOSURE SURVEY

Post-closure surveys will be needed when preparing plans for the decommissioning, decontamination and demolition of plant and buildings, as discussed in Volume II. It may be necessary to determine:

- the nature of surface deposits
- the contents of tanks, pipes, drums etc.
- the nature and extent of contamination of the building fabric.

Where investigation of the site itself, or surface deposits, is required, the procedures described in this Volume will be applicable. More specialist aspects (e.g. the investigation of the fabric of buildings for contamination) are dealt with in Section 3 of Volume II.

> **Box 2.7** *Presentation of the report*
>
> It is frequently convenient to produce the report in two volumes:
>
> 1. A 'factual' report describing the work done and the rationale ((1) to (7) in Box 2.6)
>
> 2. An 'interpretative' report containing an assessment of the results and recommendations on the actions that should follow ((8) to (13) in Box 2.6).
>
> The division of the report into volumes makes it easier to provide the interested parties with factual information without risk of prejudicing their view of the results or giving away information/opinions that may have commercial value or be material to discussions with regulatory bodies. However, all parts of the report should be properly cross-referenced.
>
> Where the report is divided in this manner, different organisations may be employed to produce the different parts. If this is the case, it is important to ensure full and unbiased liaison between the organisations, so that the overall report is comprehensive and accurate.

REFERENCES

1. BRITISH STANDARDS INSTITUTION. *Draft for development: Code of Practice for the Identification of Contaminated Land and for its Investigation.* BS DD175:1988. BSI (London), 1988

2. BRITISH STANDARDS INSTITUTION. *Code of Practice for Site Investigations.* BS 5930:1981. BSI (London), 1981

3. SITE INVESTIGATION STEERING GROUP. *Site Investigation in construction: 1. Without investigation ground is a hazard. 2. Planning Procurement and Quality management. 3. Specification for ground investigation. 4. Guidelines for the safe investigation by drilling of landfills and contaminated land.* Thomas Telford (London), 1993

4. LEACH, B. A. AND GOODGER, H. K. *Building on Derelict Land.* Special Publication 78, CIRIA (London), 1991

5. HEALTH AND SAFETY EXECUTIVE. *Evaluation and Inspection of Structures.* HS(G)58. HMSO (London), 1990

6. HOBSON, D. M. Rational site investigation. In: *Contaminated Land: Problems and Solutions.* T. Cairney (ed.) Blackie Academic (London), 1993, pp 29-67

7. NEDERLANDS NORMALISATIE-INSTITUUT. *Soil: Investigation Strategy for Exploratory Survey.* NVN 5740. NNI (Delft), 1991

8. INTERNATIONAL ORGANISATION FOR STANDARDISATION. *Soil quality – sampling, Part 5: Sampling strategies for the investigation of soil contamination of urban and industrial sites* (3rd draft). ISO CD10381. September 1993 (available from BSI)

Further reading

See Appendices 1, 2, and 3,

3 Implementation of site investigations

3.1 PROJECT MANAGEMENT

Site investigation invariably involves a number of parties, for example:

- the client

- a consultant who should, but may not, possess specialist expertise in dealing with contamination

- a specialist sub-consultant if the main consultant lacks contamination expertise

- a specialist contractor experienced in the investigation of contaminated land

- a specialist laboratory equipped to undertake appropriate chemical analyses if the ground investigation specialist contractor does not have such in-house facilities

- the regulatory and statutory bodies who are not party to the contract but whose requirements may have to be complied with under the contract

- the landowner, if not the client, and adjoining landowners/users etc.

- the public who may be affected by on-site works.

Besides the technical objectives (and requirements), planning and management will be concerned with non-technical matters including:

- responsibilities and roles of the interested parties

- legal/consent issues

- insurance requirements

- selection and procurement of suitable specialists to do the work

- specification and conditions of contract

- personnel requirements

- quality assurance and quality control

- health and safety issues

- environmental protection

- long-term requirements such as monitoring.

In the interests of proper planning and management one individual with project management responsibilities should be appointed to ensure liaison between specialists within the investigation team and with clients, regulatory bodies and other organisations external to the team.

The proper ordering and integration of different aspects of the investigation is vital. Sufficient resources, including time, should be made available to carry out the work, bearing in mind that site investigation is an iterative process and both timescales and costs may change as work progresses.

3.2 LEGAL ASPECTS

A detailed discussion of the legal requirements relating to contaminated sites is given in Volume XII. Aspects of concern in site investigation include:

- possible need for planning permission to carry out the investigation (see Section 5, Volume XII)

- the obligations of the employer to protect the health and safety of the workforce, and that of the general public (see Section 6, Volume XII). Note in particular requirements under:
  - Control of Substances Hazardous to Health (COSHH) Regulations, 1994
  - Management of Health and Safety at Work Regulations, 1992
  - The Workplace (Health, Safety and Welfare) Regulations, 1992
  - The Provision and use of Work Equipment Regulations, 1992
  - The Personal Protective Equipment at Work Regulations, 1992
  - The Construction (Design and Management) Regulations, 1994
  - Transport and packaging regulations

  and their Northern Ireland equivalents.

- occupier's liability issues including the need to protect public health, including that of trespassers (see Section 7, Volume XII)

- the need to protect environmental (air, water, land, flora and fauna, archeological remains etc.) quality (see Section 8, Volume XII)

- the need to comply with legal requirements regarding the on-or off-site disposal of wastes (see Section 8, Volume XII).

- access and rights of way.

3.3 PROCUREMENT

Safe, effective and economic site investigation depends on making the right procurement decisions[1, 2]. Assuming appropriate expertise is not available internally, clients will normally need to procure both specialist advice and contracting services. Two main options are available[3]:

1. Use of a professional adviser, with the separate employment of a contractor(s) for physical work, testing and reporting as required.

2. Use of a single contract covering specialist advice, physical investigation, testing and reporting.

The selection of specialist(s) having the necessary technical competencies and experience in this type of work is crucial to a successful outcome. All team members should understand the objectives and background to the work, their own roles and responsibilities, and how the different specialist inputs are to be coordinated.

A number of interrelated issues must be settled prior to the actual procurement of investigation work:

- source of expert advice – appointment of a specialist consultant/adviser

- whether to employ a combination of consultant and specialist contractor(s) (e.g. on-site works, chemical analysis, geotechnical testing, ecological assessment) or a single specialist organisation providing all services

- method of procurement — selective tendering or negotiated contract
- contract arrangements — standard or specific.

The method of procurement is often determined by client requirements, such as public accountability. However, the early appointment of a suitably experienced consultant/adviser to manage, procure and implement the investigation (in addition to the initial planning and design, and subsequent assessment and evaluation) is strongly recommended in the interests of achieving an economic and technically sufficient investigation. Note that some specialist organisations offer design, implementation, analytical services, assessment and supervision of remedial works as a complete package and the use of such organisations may be appropriate in certain applications. More detailed guidance on the selection of appropriate specialists for investigation work is given in Section 3.5.

Investigation work is normally procured[3] either through selective tendering (on a Specification and Bill of Quantities basis), or by contract negotiation with one or more organisations. Both approaches have advantages and disadvantages; neither guarantees that an organisation with adequate expertise will always be selected.

Whatever form of procurement is adopted the first task is to prepare appropriate documentation defining the nature of the work required, and the duties and obligations of the various parties. Although many site investigations proceed smoothly without formal contract documentation, protracted and expensive delays can occur if a dispute arises and a contract is not available. The principal purpose of a contract therefore is to ensure that appropriate procedures are agreed in advance of undertaking the site investigation, in the event that the work does not proceed to the mutual satisfaction of the parties involved.

Standard forms of contract, such as the Institution of Civil Engineers Conditions of Contract for Ground Investigation 1983[4] may be used, although these have not been prepared specifically for the investigation of contaminated sites. All standard clauses should be reviewed and modified where necessary to ensure adequate provision for items such as health and safety, environmental protection and waste disposal (see also Section 3.4).

Remuneration is a key issue in procurement. Competitive tendering, either for specialist consultancy or contracting services, is the current norm for many projects. This has the potential disadvantage that costs can assume greater importance than technical sufficiency, but it can be operated successfully provided the tender list is restricted to a few firms able to tender on an equal footing. The Site Investigation Steering Group[1] recommends a select tender list of no more than three firms of equal standing. To achieve a select list, careful consideration needs to be given to expertise, relevant experience and available resources (see Section 3.5).

The most common form of payment for contracting services is through the ad measurement contract, using a Bill of Quantities of agreed rates (see Volume XI for more detailed discussion), although lump sum payment is required on some projects by certain clients. Lump sum contracts have obvious attractions for clients and may be appropriate for small-scale and well defined investigation projects. However, they are not well suited to the phased approach to investigation recommended here since the scope of the work cannot easily be amended in the light of the conditions actually found. Adoption of a lump sum contract will usually place a constraint on the investigation, rendering it less effective than it might otherwise be. The same applies to specialist consultancy, where financial constraints rarely produce the best quality advice. The Site Investigation Steering Group[1] recommends that remuneration of an adviser should be on a time basis, using agreed rates and expenses. For both advisory and contracting services, however, the client should expect to be able to agree a programme and budget for the work, and restrictions (e.g. prior approval) on any work done beyond authorised levels.

A detailed discussion on procurement is given in Volume XI; further guidance can be obtained from other recent CIRIA publications.[3, 5]

3.4 SPECIFICATION

In addition to BS DD 175[6], BS 5930[7], the ISO Standards and other documents (see Appendix 1), reference should be made to current UK specifications for Ground Investigation, (see Table 3.1). Of these, only the recently revised Specification for Ground Investigation[8] includes clauses specific to contaminated land (note this document is intended to replace not only the old ICE document[9] but also the DoT specification[10]). Only limited guidance is given; for example little advice is given on sampling protocols, or analytical and testing strategies specific to contamination. However, the ICE Specification is the most recent consensus view on ground investigation specification in the UK and it provides the facility, though the 'Particular Specification' section, to document any site-specific requirements and constraints relating to contamination, or any other matter. Section 4 and the Appendices provide more detailed guidance that can be used to draft the 'Particular Specification' section for individual projects.

Table 3.1 *Specifications for ground investigation*

Authority	Title	Reference
ICE	Specification for Ground Investigation 1989	9
ICE	Specification for Ground Investigation 1993	8
Department of Transport	Specification and Method of Measurement for Ground Investigation 1987	10
AGIS	Specification for Ground Investigation 1979	11

Care must always be exercised when adopting specifications prepared for geotechnical investigation purposes to ensure that detailed requirements (e.g. on the formation of boreholes or trial excavations; the taking, handling and preservation of samples) are appropriate to the task in hand.

The need for care to is shown by the fact that the advice in the ICE Specification[8] on the preservation of samples (to be kept between 2 and 45°C) contradicts good practice in respect of environmental samples — samples should be kept at less than 4°C. Note also that the recommended analytical methods for contaminated soils, the arbitrary division of contaminants into primary and secondary, and the inclusion of 'dependent' analytical options, have not been endorsed by the relevant professional bodies (e.g. the Analytical Division of the Royal Society of Chemistry). As noted in Section 3.6.3, a series of analytical methods specific for the purpose is being produced by the International Organisation for Standardisation (ISO). These will be adopted as British Standards and should be the preferred choice.

Specifications for the investigation of contaminated land, and related aspects such as sampling strategies, have been prepared by European and North American organisations (see Appendix 1). These can be useful when drafting technical specifications for work in the UK provided due account is taken of the (sometimes substantial) differences in approach to ground investigation techniques and equipment that have developed in these countries as a result of historical and geological factors.

Because of the current lack of suitable (dedicated) technical specifications for the investigation of contaminated land in the UK, appropriate clauses for most, if not all, investigations will have to be drafted on a site-specific basis. Appropriately experienced personnel must be used for such work. Organisations able to demonstrate a successful track record in relevant, comparable projects should be in a position to prepare specifications that satisfy the technical, legislative and regulatory demands of the work, on the basis of current good practice and the existing knowledge base.

3.5 SELECTION OF APPROPRIATE SPECIALISTS

All ground investigation work is specialised and should be carried out only by organisations having the necessary specialist expertise, personnel and equipment. The same principles apply to the investigation of contaminated land, and to all potential parties, including the site investigation contractor, the analytical laboratory and the organisation offering consultancy advice.

The investigation of a contaminated site differs from other forms of investigation in a number of ways including:

- The inherent complexity and diversity of the problems affecting many contaminated sites. This restricts the scope for a standardised approach and demands a high level of experienced, professional judgement in the design of site investigation strategies.

- The specific technical requirements of investigation. Poorly located sampling positions, incorrect handling of samples or inappropriate methods of analysis can at best mislead and at worst endanger health and safety, and the environment.

- The need for experienced personnel in the field and laboratory who are familiar with the instruments and techniques to be employed.

- The need for site personnel who are familiar with, and willing to observe, strict health and safety, and environmental protection policies.

In practice, the selection of consultants and/or contractors often relies on word-of-mouth recommendations, perhaps via local authority select lists, or the identification of prominent individuals or organisations from published literature, seminars and directories (see references 12 to 16 and Appendix 4). Note that the Association of Environmental Consultancies (AEC) has introduced a scheme for AEC registration of members providing specialist contaminated land services.

The *Site investigation in construction* document[1] provide advice on the selection of the geotechnical team and make recommendations about the client's obligations in this respect.

There are also systematic checks that can be made. Clients should consider whether organisations can show a detailed understanding and awareness of the full range of technical and managerial issues likely to be involved; organisations should be able to offer a track record that demonstrates the successful outcome of similar, or more complex, projects of the type under consideration. Box 3.1 lists factors to be considered when appointing specialist advisers and/or contractors; Appendix 4 lists some points to be taken into account in the selection of site personnel engaged in sampling.

There are no 'hard and fast' rules for the type and size of team that should be employed by a client to carry out the investigation. It is conventional in construction projects to employ a consultant to advise on the scope and extent of the investigation, to undertake the appointment and direction of a specialist contractor (appointed either through competitive tendering or negotiation) and to carry out the risk assessment and selection of the remedial techniques. Such a consultant might have a chemical, biological, geological, hydrogeological or geotechnical background; however, provided their expertise and experience are relevant their background qualification may be of secondary importance. Conversely, a specialist environmental consultancy or contractor may be appointed to carry out all or part of the work.

For geotechnical investigations the Site Investigation Steering Group[1] recommends the appointment of a 'Geotechnical Adviser' for each investigation. A geotechnical adviser is defined as having chartered status and at least five years practice as a Geotechnical Specialist, a term also defined. Although these guidelines do not refer specifically to the investigation of contaminated sites they are indicative of an appropriate level of experience, for example:

- A specialist adviser for contamination investigations should have an appropriate professional qualification and at least five years experience in the investigation, assessment and remediation of contaminated land.

The Association of Environmental Consultancies has introduced a Code of Practice for members specialising in contaminated land work which will provide guidance on the experience and expertise required of project managers etc.[17]

Particular attention should be paid to the experience of individuals placed in positions of responsibility. For example a geotechnical specialist placed in charge of a field investigation may not have appropriate experience in the investigation and remediation of contaminated land; an analytical chemist or environmental scientist may not have the necessary expertise to interpret geological or hydrological conditions, or the practical aspects of site remediation.

NAMAS (National Measurements Accreditation Service) accreditation is an important factor in the selection of an analytical laboratory. Note that NAMAS accredits performance in undertaking published analytical procedures; it does not guarantee that the analytical method used is the most appropriate for a particular application.

Analytical reliability can also be gauged by performance in independently controlled inter-laboratory testing (see Section 3.6.3). The Laboratory of the Government Chemist (LGC) has recently launched a scheme (CONTEST) under the Department of Trade and Industry's 'Validity of Analytical Measurements' initiative which should lead to the introduction of appropriate performance testing schemes for contaminated soils. Some laboratories already participate in inter-laboratory testing programmes for the analysis of water samples and asbestos where standard methods are more established. Participation in such schemes should be explored when considering the use of a laboratory for the first time.

Enquiries should be made regarding the professional indemnity insurance held by the organisation(s) offering to carry out the work: organisations or individuals (e.g. a geotechnical engineer advising on human health risks) going beyond their established area of expertise may not be covered. The current uncertainty in the insurance market concerning contamination and pollution matters should also be borne in mind. There has been recent evidence of insurers withdrawing cover (particularly for gradual 'pollution' incidents) irrespective of the expertise or competence of the organisations seeking cover. At the time of writing, the current and future availability of professional indemnity insurance for contaminated land work is unclear. Since such insurance operates on a claims made basis, reliance on the professionals' insurance may not provide the guarantees required in future years.

> **Box 3.1** *Selection of appropriate specialists*
>
> Specialist advisers should:
>
> - operate a quality management system
>
> - be knowledgeable about and have expertise in risk assessment (including risk estimation and evaluation), health and safety aspects of investigation, soil and water quality standards, appropriate modelling techniques, remediation techniques, analytical and testing aspects of investigation, environmental legislation, contract procedures and validation techniques
>
> - have experience of the execution, specification and supervision of contaminated land investigations
>
> - be familiar with published guidance, e.g. from the Department of the Environment
>
> - be able to identify situations where additional specialist services or advice may be required.
>
> *Note: It is unlikely that an individual will have detailed expertise in all subjects or activities which need to be addressed in a risk assessment: for complex projects it will often be necessary to employ an organisation or team of specialists offering the breadth of services and expertise required.*
>
> Specialist service providers (contractors, laboratories) should:
>
> - operate appropriate quality management systems
>
> - have third party accreditation for testing procedures etc.
>
> - have experience of similar work
>
> - employ appropriately qualified and experienced key senior staff
>
> - ensure field staff are appropriately qualified and receive specialist training
>
> - employ competent staff certified to a recognised standard (e.g. drillers under the British Drillers Association scheme[18])
>
> - be able to demonstrate adherence to good practice in terms of health and safety
>
> - be able to demonstrate familiarity with relevant legislation particularly on health and safety, environmental protection, and waste disposal
>
> - have the resources (manpower and equipment) to carry out the necessary work

3.6 QUALITY MANAGEMENT

3.6.1 General requirements

The quality of site investigation work is an important issue for all the parties involved. The client's quality objectives are to achieve technically sufficient investigation work, at the right time and at the right price. Society requires that due attention is paid to matters such as health and safety and environmental protection. The organisation carrying out the work will wish to meet all the client's requirements and make a profit. Contractors and subcontractors require clear instructions and information in a timely fashion.

Quality is defined in BS 4778 : Part I[19] and is discussed in more detail in Volume XI (Planning and management). The definition of quality in a site-specific context may be somewhat subjective; work done by one organisation in one situation may be considered to be of poor quality by another organisation having greater experience or expertise in a particular area. It is generally easier to judge performance if, at the outset, the objectives and requirements of the work and the procedures to be followed have been specified.

In a site investigation context, achieving quality generally requires:

- a clear statement of objectives
- clear assignment of responsibilities
- procurement of appropriate (specialist) expertise
- development of technical specifications
- development of methods for monitoring the quality of the work done and rectifying poor quality work
- good communication (both internally and externally).

Some key aspects of the site investigation process which should be subject to a comprehensive system of progress and performance checks are listed in Box 3.2.

Both the client and organisations providing services have responsibilities in ensuring the quality of site investigation work.

The client has a responsibility to ensure that the work is properly conceived in the first place and that it is properly carried out. It is important to recognise that site investigation is a continuous process with sequential tasks. If any component fails, the process also fails. In this context the client is responsible for managing the project relationships and ensuring the competence of each participating organisation.

Box 3.2 *Key aspects of site investigation which should be subject to performance monitoring*

Compliance with all relevant legal requirements

Review of documentary evidence during desk study

Location and recording of observations during site reconnaissance

Procedures used to identify potential hazard/pathway/target combinations and to select 'plausible' scenarios for further assessment

Siting and installation of exploratory excavations

Establishment and performance of environmental protection measures

Waste disposal arrangements (Duty of Care requirements etc.)

Implementation of health and safety procedures

Collection and handling of samples

Storage and preparation of samples

Methods of analysis and testing

On-site recording protocols

Reporting of data

Reporting procedures used in estimation of risks

Input to, and use of any models used to aid interpretation of the data

Participation by contracting parties in Accreditation schemes for Quality Management Systems (e.g. for BS 5750) and analytical and testing services (e.g. NAMAS)

Organisations offering site investigation services should make sure that the client is not given a false impression of what can be achieved by a site investigation/risk assessment, and is told at the outset of any uncertainties that may remain on completion of the work. Limitations imposed, for example by time and cost constraints, should be spelled out; an investigation may be of very high quality in terms of execution in the field but appear to be of low quality because of inherent limitations in the scope and amount of work carried out.

The quality management systems used by consultants and contractors should be open to independent inspection. They need not be the subject of third party accreditation (e.g. to BS EN ISO 9000 (formerly BS 5750)[20] although this is increasingly seen as desirable and is sometimes made a condition of contract. Note that independent accreditation implies a capability to achieve quality: it does not guarantee that the requisite quality will be achieved in any individual project. In the absence of third party accreditation, the client may properly require confirmation that an internal system of a comparable standard is operated. The Association of Geotechnical Specialists has published guidance[21] on the application of quality management systems to geotechnical investigations.

CIRIA has been involved in providing guidance about quality assurance within the construction industry for a number of years and has issued a number of relevant publications, including a client's guide to quality assurance for construction and a guide to the interpretation of BS 5750 in the construction industry[22-27]. There is no formally approved (i.e. by the National Accreditation Council for Certification Bodies) accreditation scheme specific to the investigation of contaminated land.

3.6.2 Site operations

Site operations which should be subject to quality assurance and quality control procedures include:

- exploratory hole formation/monitoring point installation
- sample collection and handling
- waste disposal
- environmental protection
- health and safety
- monitoring.

All aspects of the installation and development of trial excavations, boreholes and monitoring wells etc., should be subject to detailed technical specification and the application of quality assurance procedures to ensure adequate standards of workmanship.

Special consideration should be given to controlling the quality of groundwater monitoring installations in relation to:

- materials and methods of construction
- construction details
- preparation and handling of materials
- use of on-going checks to measure standards achieved (e.g. introduction of gravel pack, introduction of seals)
- general standards of workmanship and cleanliness (e.g. cleaning drilling equipment, care in selection of flushing media)

- well development procedures
- post-treatment monitoring.

Various physical tests (pumping tests, levels monitoring etc.) can determine whether the installation is performing as required; up-gradient and down-gradient test data can be used to indicate the presence of contamination which cannot be attributed to the source being monitored.

3.6.3 Chemical analysis

Quality assurance (QA) in chemical and other forms of analysis defines the acceptance limits of measurement and monitoring data in terms of sensitivity, reproducibility, detection limits and accuracy. Quality assurance/quality control protocols are employed to demonstrate the reliability of sample analysis data and compliance with documented sample handling, storage, preparation and test methods.

When a client invites a test laboratory to (collect and) test samples, it is expected that the samples collected on site are representative of the material being sampled, and the results obtained are:

- accurate
- precise
- reproducible (inter and intra-laboratory)
- reliable
- timely.

The main source of guidance on the accreditation of laboratories is ISO Guide 25: *General requirements for the technical competence of testing laboratories*[28]. This is being incorporated in Europe in EN45001, *General criteria for the operation of testing laboratories*[29]. In the UK the standard is implemented by NAMAS. Eighteen other European countries have implemented similar schemes and are working towards mutual recognition. There are fifteen elements to the NAMAS accreditation standard: key elements are set out in Appendix 5. In practice only a third of these can be satisfied on a routine basis for soil samples.

Agreements have been reached with other non-testing assessment schemes. All requirements of sub-contractor/supplier assessments included in BS 5750/ISO 9000, Lloyds Register, and Det norske Veritas (DnV) registration are deemed to be satisfied (in terms of testing) through accreditation by NAMAS. When a NAMAS accredited laboratory has been selected the client has the right to expect documented evidence that:

- the integrity of the sample has been maintained
- measurements have been carried out using approved standard methods
- a known level of accuracy and precision can be attached to each result
- an 'identical' result would be obtained if the sample were to be retested, either by the same, or another accredited laboratory.

'Standard' methods are available for water analysis, and are currently under development for soils by the International Standards Organisation (see Appendix 7). Until standard methods for soils are available, inter-laboratory proficiency tests cannot readily be carried out. Standardised test methods are critical in soil analysis because most of the methods in current use are empirical, i.e. they do not measure the true 'total' concentrations in the sample of the substance tested for but only that portion which is soluble in the chosen extractant under the test conditions (even though the values are frequently reported as total concentrations, see Table 3.2). The results obtained may also depend on the method used for the actual determination.

Table 3.2 *Example of measurement of 'total' metals by digestion with nitric/hydrochloric acids, followed by measurement by atomic absorption (AA) or inductively coupled plasma (ICP) spectrometry*

Determinand	Recovery (%)
Aluminium	40
Cadmium	70-90
Calcium	50
Chromium	70-90
Copper	70-90
Iron	70-90
Lead	70-90
Magnesium	70
Manganese	70-90
Potassium	20
Sodium	4
Zinc	70-90

In the absence of standard methods and given the shortage of appropriate proficiency testing programmes the client needs to know:

- what to look for in a laboratory in addition to accreditation
- how to verify adequate performance.

The first can be resolved by evaluating the quality management arrangements in operation at prospective laboratories including both the quality control system and performance in proficiency test programmes where these are relevant. One appropriate Quality Control system has been proposed by the Water Research Centre[30]. In the UK, the main inter-laboratory proficiency programme is 'Aquacheck', run by the Water Research Centre, which has recently been extended to cover 'trace elements in soils' (see also the LGC CONTEST noted above). The use of certified materials can also assist in checking for bias.

Clients have two ways of verifying the results obtained on their samples; both involve sub-dividing the sample and this can introduce additional errors. Duplicate samples may be submitted or sub-samples can be submitted to other laboratories. In the latter case, there will then be the difficulty of explaining any significantly different results. Clients may also consider submitting their own standard samples.

Note that the greatest source of variation in soil analysis derives from soil sampling procedures and that results may subsequently be affected by handling and containment. Therefore these procedures should be standardised as far as possible.

Representative elements of a QA programme for contaminated site samples include:

1. *Sampling*: Adequate records of sampling and sample numbers, sampling techniques compatible with contaminants and analytical methods to be employed, possible preservation of 'archive' samples.

2. *Sample handling*: Use of appropriate containers and storage conditions.

3. *Sample preparation*: Homogeneity/heterogeneity of sample, sub-sampling procedures, use of preparation methods that will not affect result, recording of procedures.

4. *Instruments and techniques*: Calibration, testing of reproducibility on split samples, running standard samples.

Field instruments and test kits should be calibrated in the laboratory and/or tested under controlled conditions to determine analytical characteristics (accuracy, reproducibility etc.). Calibration standards should be employed in the field where appropriate.

3.7 HEALTH AND SAFETY

The investigation of contaminated sites, by definition, will involve investigation personnel in close, and possibly prolonged contact with potentially hazardous materials. BS DD175:1988[6] states:

> 'Prior to sampling, it is essential to consider the risks to the health and safety of the investigators and to other persons and property and to take the appropriate precautions. The main responsibility rests with the leader of the investigation team, who should have specialist knowledge and experience and be familiar with the purpose of the investigation. Each member of the team should have defined responsibilities, and be given clear briefings and instructions before work starts.'

Detailed guidance on health and safety provision in the investigation of contaminated sites is given in a CIRIA Special Publication[31], in a Health and Safety Executive document[32] and other related literature (see for example reference 33 and British Drilling Association guidance on safe drilling practices[34,35], and references 36 to 39).

Site investigation staff should be suitably trained and equipped. Some measures for protection, monitoring and control are listed in Table 3.3. A safety checklist is provided in Box 3.3.

Table 3.3 *Health and safety measures that may be required for site investigations*

Protective clothing and equipment	Monitoring equipment	Safety procedures
Overalls, boots, gloves and helmets	Hand-held gas monitors	Training
Eye protection	Automatic gas detectors	Permit to work systems
Face masks and filters	Personal monitors	Notification to emergency services
Breathing apparatus	Environmental monitoring equipment	Access to telephone contact
Safety harness and lanyards		Decontamination facilities for plant
Safety torches		Decontamination facilities for personnel
Fire extinguishers		Safe sampling procedures
First aid equipment		Safe sample handling procedures
		Access for emergency vehicles

Box 3.3 *Safety checklist*

1. Has a COSHH assessment for all hazardous substances likely to be found on the site been carried out?
2. Has using a non-intrusive investigation method or one which limits ground disturbance to minimise contact between employees and contaminants been considered?
3. Have sample locations taken account of potentially most hazardous areas?
4. From the COSHH assessment which of the following equipment will be required to help ensure health and safety on site?
   - safety foot wear?
   - gloves?
   - overalls?
   - eye protection?
   - gas monitors/detection instruments?
   - respiratory protective equipment?
   - safety harnesses and man-lifting equipment?
   - personal exposure detectors?
   - first aid box/first aid room?
   - decontamination facilities?
   - washing facilities?
   - vehicle washing/decontamination facilities?
   - contaminated equipment disposal facilities?
   - separate eating/resting area?
   - communication equipment?
5. Which of the following safety procedures have been adopted for work on the contaminated site
   - personnel briefed thoroughly on the hazards they may encounter and the appropriate precautions to be taken?
   - training in use of personal protective equipment provided?
   - staff trained to recognise unusual material, drums etc., and the necessary steps to follow?
   - no smoking, eating or drinking on the site?
   - the positioning of staff and plant upwind of excavation/boreholes?
   - Permit-to-work system for work in confined spaces or or where toxic atmospheres are present?
   - TV monitoring/recording of particularly hazardous operations?
   - monitoring for explosive gases?
   - monitoring for toxic fumes/gases in boreholes and excavations?
   - safe storage and disposal of material excavated from the site?
   - safe handling, storage and transportation of contaminated samples, taking into account their respective hazards?
   - correct labelling of sample containers, detailing potential hazards?
   - removal and decontamination of all protective clothing before going off-site?
6. Has the Managers' Guidance Chart (*see source reference*) been completed?
7. In case of emergency, are the locations of the following emergency services known and displayed appropriately?
   - local hospital with casualty ward?
   - location of nearest telephone to call fire services, police etc.?
8. Have the emergency services been informed/consulted?

From reference 31

Before undertaking any form of investigation on a site where contamination is suspected a COSHH assessment must be carried out. The COSHH regulations (see Section 3.2 and Volume XII) require the employer to adhere to the following principles.

- Avoid exposure to substances hazardous to health, and only when this is not reasonably practicable should methods of controlling exposure be employed.

- If exposure has to be controlled, personal protective equipment should only be used as a last resort i.e. other forms of control such as ventilation should always be considered first.

Where site reconnaissance forms part of the preliminary investigation, the COSHH assessment should be based on the results of the desk study. It may be possible to refine the assessment once the preliminary investigation is completed. It should be kept under review as the investigation proceeds. The HSE has issued specific guidance in relation to peripatetic workers.[40]

Health and safety should be a fundamental consideration in the design and the selection of investigation methods.[38] Therefore ask, for example,

- if is it really necessary to disturb this (part of the) site?

- whether investigation of this area should be delayed until investigations elsewhere have provided a better picture of the nature and extent of contamination?

At least initially, investigation techniques (e.g. driven probes) can be employed that minimise the exposure of the investigation team to contaminants. In situations where extremely hazardous, reactive or volatile contaminants are expected, consideration should be given as to whether trial pits and other open excavations (which may present a risk to both investigation personnel and neighbouring populations) are the best way of exploring a site. The phasing of investigations can make an important contribution to safety. For example, where either substantial gas or leachate migration is suspected, initial investigation should be concentrated on the boundaries of the site. Direct interference with the source at this stage could result in an uncontrolled release of toxic, flammable or strongly smelling gases and vapours. When the results of these exploratory investigations are available, subsequent investigations can employ methods that allow control over releases to the environment and minimise the chances of exposure of the work force.

The HSE advises that where possible details of contamination and outline precautions should be included within tender documents for site investigation contracts[32]. Employers (e.g. consultants/contractors) should be asked to provide a COSHH assessment and a general health and safety statement when placing a bid for work on a contaminated site.
It is recommended that:

1. An individual (safety officer) responsible for health and safety issues is appointed for each investigation project. Preferably this person should not have other responsibilities, because of the potential for conflict, although it is recognised that on small projects, at least, this may be unavoidable. The importance of health and safety issues should be recognised by all personnel involved.

2. The Health and Safety Executive is be consulted on the precautions to be taken during the investigation of contaminated sites — such consultation should take place after the findings of the preliminary investigation have been assessed so that site-specific information can be taken into account.

3. The local Environmental Health Department is consulted on measures to protect people living in the vicinity of the site.

The Construction (Design and Management) Regulations 1994 require the development of a health and safety plan for all construction sites and the appointment of a planning supervisor and a principal contractor to coordinate health and safety considerations during the different stages of a project. The Regulations took effect early in 1995.

3.8 ENVIRONMENTAL PROTECTION REQUIREMENTS

Since the investigation of contaminated sites involves physical disturbance, there is a possibility of the uncontrolled release of potentially hazardous materials to the environment. Examples of protective measures are listed in Box 3.4.

Box 3.4 *Examples of environmental protection measures during site investigation*

Controlled disposal (e.g. at a treatment centre, authorised disposal site) of surplus materials from exploratory excavations

Protection of uncontaminated lower strata, and groundwater, during drilling operations in areas of confined waste (e.g. gasholder base, landfill) or in open ground

Cleaning site vehicles and controlling vehicle movement on site

Protection of surface water bodies during excavation activities (e.g. using temporary drainage) and controlled disposal of effluents

Control over releases of odours and toxic gases/ vapours (e.g. shrouds fitted to boring equipment; full controlled ventilation may be required in extreme circumstances)

Protection of ground surface (e.g. using temporary liners etc.) from extracted solids and liquids during drilling

---

3.9 LONG-TERM SAMPLING AND OFF SITE WORKS

There may be a need to undertake long-term sampling or monitoring, both on and off the site. A wide range of sampling activities may be required including the collection of surface water, vegetation and soil samples. For hydrological investigations, monitoring installations may have to be placed in off-site locations. Appropriate approvals to enter adjoining land or other premises for sampling purposes, and agreement for long-term access, should be obtained well in advance of the need to commence sampling.

Provision must be made in the scheduling of the site investigation for the application for, and granting of, approvals to undertake off-site works.

REFERENCES

1. SITE INVESTIGATION STEERING GROUP
   Site investigation in construction:
   1. Without investigation ground is a hazard
   2. Planning procurement and quality management
   3. Specification for ground investigation
   4. Guidance for the safe investigation by drilling of landfills and contaminated land
   Thomas Telford (London) 1993

2. INSTITUTION OF CIVIL ENGINEERS. *Inadequate Site Investigation.* Thomas Telford (London), 1991

3. UFF, J.F. and CLAYTON, C.R.I. *Recommendations for the procurement of ground investigation*. Special Publication 45. CIRIA (London), 1986

4. INSTITUTION OF CIVIL ENGINEERS. *ICE Conditions of Contract for Ground Investigation*. Thomas Telford (London), 1983

5. UFF, J.F. and CLAYTON, C.R.I. *Role and Responsibility in Site Investigation*. Special Publication 73. CIRIA (London), 1991

6. BRITISH STANDARDS INSTITUTION. *Code of Practice for the Identification of Contaminated Land and for its Investigation*. DD175:1988 Draft for Development. BSI (London), 1988

7. BRITISH STANDARDS INSTITUTION. *Code of Practice for Ground Investigation*. BS 5930:1981. BSI (London), 1981

8. SITE INVESTIGATION STEERING GROUP. *Site investigation in construction: 3. Specification for ground investigation*. Thomas Telford (London), 1989

9. INSTITUTION OF CIVIL ENGINEERS. *Specification for ground investigation*. Thomas Telford (London), 1989

10. *Specification and method of measurement for ground investigation*. HMSO (London), 1987

    (Note: issued in the names of the Department of Transport, Department of the Environment, Property Services Agency, Scottish Development Department, Welsh Office and the Department of the Environment for Northern Ireland)

11. ASSOCIATION OF GROUND INVESTIGATION SPECIALISTS. Specification for Ground Investigations. *Ground Engineering*, 12 (5), 1979

12. THE BRITISH GEOTECHNICAL SOCIETY. *The Geotechnical Directory of the United Kingdom*. Third edition
BGS (London), 1992

13. ROYAL SOCIETY OF CHEMISTS INSTITUTION OF ENGINEERS INSTITUTION OF BIOLOGISTS. *List of contaminated land specialists*. RSC (London)

14. ENDS. *Directory of Environmental Consultants*. Third edition (1992/93)
Environmental Data Services (London), 1992

15. THE GEOLOGICAL SOCIETY. *The Geologist's Directory*. Seventh edition. The Geological Society (London), 1994

16. NATIONAL PHYSICAL LABORATORY. *NAMAS Directory of accredited laboratories*. NPL (Teddington), 1993

17. ASSOCIATION OF ENVIRONMENTAL CONSULTANCIES. *Code of Practice (Contaminated Land)*. AEC, 1994

18. BRITISH DRILLING ASSOCIATION LTD. *Ground investigation drillers' accreditation scheme*. BDA (Brentwood)

19. BRITISH STANDARDS INSTITUTION. *Quality Vocabulary*. BS 4778 BSI (London), 1987 and 1991

20. BRITISH STANDARDS INSTITUTION. *Quality Systems*. BS EN ISO 9000: 1994 BSI (London), 1994

21. ASSOCIATION OF GEOTECHNICAL SPECIALISTS. *Quality management in geotechnical engineering, a practical approach*. AGS, 1990

22. ANON. *National quality assurance forum for construction*. Special Publication 61. CIRIA (London), 1988

23. ANON. *Client's guide to quality assurance in construction*. Special Publication 55. CIRIA (London), 1988

24. OLIVER, G.B.M. *Quality management in construction: interpretations of BS 5750 (1987) – 'Quality systems' for the construction industry*. Special Publication 74. CIRIA (London), 1990

25. OLIVER, G.B.M. *Quality management in construction: implementation in design services organisations*. Special Publication 88. CIRIA (London), 1992

26. SADGROVE, B.M. *Quality assurance in construction – the present position*. Special Publication 49. CIRIA (London), 1986

27. BARBER, J.N. *Quality management in construction: contractual aspects*. Special Publication 84. CIRIA (London), 1992

28. INTERNATIONAL ORGANISATION FOR STANDARDISATION/BSI. BS 7502:1989/EN 45002 *General Criteria for the Assessment of Testing Laboratories*. Guide 25. BSI (London), 1989

29. INTERNATIONAL ORGANISATION FOR STANDARDISATION/BSI. *BS 7501:1989/EN 45001: General Criteria for the Operation of Testing Laboratories*. BSI (London), 1989

30. CHEESEMAN, R.V. and WILSON, A.L. (revised by GARDNER, M.J.). *A manual for analytical control for the water industry*. Water Research Centre (Medmenham), 1989

31. STEEDS, J.E., SHEPHERD, E. and BARRY, D.L. *A guide to safe working practices for contaminated sites*. Report 132 CIRIA (London), in the press

32. HEALTH AND SAFETY EXECUTIVE. *Protection of Personnel and the General Public during Development of Contaminated Land*. HS(G)66. HMSO (London), 1991

33. SITE INVESTIGATION STEERING GROUP. *Site investigation in construction: 4. Guidelines for the safe investigation by drilling of landfills and contaminated land*. Thomas Telford (London), 1993

34. BRITISH DRILLING ASSOCIATION. *Code of safe drilling practice, Part 1: surface drilling*. BDA (Brentwood), 1981

35. BRITISH DRILLING ASSOCIATION. *Guidelines for the drilling of landfill, contaminated land and adjacent areas*. BDA (Brentwood), 1991

36. US DEPARTMENT OF HEALTH AND HUMAN SERVICES. *Occupational safety and health guidance manual for hazardous waste site activities*. National Institute for Occupational Safety and Health (Cincinnati), 1985

37. US ENVIRONMENTAL PROTECTION AGENCY. *Standard Operating Safety Guides*. USEPA (Washington D C), 1988

38. SMITH, M. A. Safety aspects of waste disposal to landfill. In: *Proceedings of a Conference on Planning and Engineering of Landfills*. Midlands Geotechnical Society (Birmingham), 1992, pp 9-21

39. O'BRIEN, A. A., STEEDS, J. E. and LAW, G. A. Case study: Investigation of Long Cross and Barracks Lane landfill sites. In: *Proceedings of a Conference on Planning and Engineering of Landfills*. Midlands Geotechnical Society (Birmingham), 1992, pp 31-34

40. HEALTH AND SAFETY EXECUTIVE. *COSHH and Peripatetic Workers*. HS(G)77. HMSO (London), 1992

4 Sampling and testing

4.1 DEVELOPING A STRATEGY

Five main questions have to be addressed:

1. What type of samples should be collected ?
2. Where should samples be collected ?
3. How many (or how often and for how long) should samples be collected ?
4. How should the samples be collected ?
5. How should the samples be analysed or tested ?

The answers to these questions depend on :

- the medium to be investigated (e.g. soil, water, ambient air, soil gas, flora and fauna etc.)
- the phase of investigation (e.g. preliminary, detailed, supplementary or for compliance/performance purposes)
- how the test and analytical data are to be used (e.g. to characterise hazards and targets, to determine compliance, to determine baseline conditions in a surface or groundwater body)
- the duration of sampling (short or long-term)
- the characteristics to be determined (e.g. concentration, volatility, solubility, pathogenicity, radioactivity, particle sizes, permeability etc.)
- the amount and quality of data available already (e.g. findings of preliminary investigation, results of pilot-scale studies etc.)
- whether the investigation for hazard and risk assessment is to be integrated with investigations for other purposes (e.g. construction).

There is no universally applicable approach to sampling and testing; appropriate strategies must be developed for each investigation taking into account site-specific objectives and conditions.

Health and safety and environmental considerations may affect the scope and type of sampling and testing carried out; for example, during exploratory investigations and the early stages of a detailed investigation, it may be necessary to use sampling techniques that minimise disturbance to the site (see Section 3). Other potential constraints on the sampling and testing strategy (see Section 3) include restricted physical access (due to operational buildings or services, concrete or other hardstanding, lack of ownership/authorisation to enter the property) and limited time or budget.

Sampling and testing may be two separate activities i.e. collection of a sample in the field followed by testing (including chemical/physical/biological analyses) in an off/on site laboratory. Field testing (e.g. on-site measurements of soil gases, water or air quality) is a combination of the two activities and both sampling and analytical aspects must be addressed at the same time.

4.2 SAMPLING STRATEGIES

4.2.1 Design aspects

Table 4.1 lists the factors which should be addressed when designing sampling strategies.

Table 4.1 *Design aspects of sampling*

Aspect	Variable
Sampling strategies	Position and types of exploratory excavations Position and types of samples Numbers of samples Frequency of sampling Duration of sampling Use of modelling techniques
Sampling procedures for :	Soils, made-ground, sediments, sludges, surface water, groundwater, drummed material, flora and fauna etc.
Phased sampling	Preliminary sampling Detailed sampling Use of retained sampling

Setting out detailed hypotheses as to the nature and distribution of the contamination can help the investigator make sensible decisions on the design of the sampling strategy before commencing the investigation; it also makes it easier to determine afterwards whether the strategy adopted was reasonable give the understanding the investigator had about contamination at the time.

Hypotheses are particularly useful during the exploratory phase where the aim is to carry out a limited amount of work to confirm the presence of contamination (and anticipated hydrological or geological conditions), or to prepare for a fuller and more detailed investigation. Five principal hypotheses might apply (see Box 4.1). Where appropriate, the site may be divided into a number of zones and separate hypotheses (and hence sampling strategies) developed for each.

Box 4.1 *Typical hypotheses*

H1   The site (or defined part of it) is uncontaminated

H2   Contamination is heterogeneously distributed with known point source(s)

H3   Contamination is homogeneously distributed

H4   Contamination is heterogeneously distributed with unknown point source(s)

H5   Contamination is uniformly heterogeneous (i.e. shows great variations in both horizontal and vertical planes with little discernible rational pattern)

The potential for contaminant migration into or out of the site should be addressed.

4.2.2 Sampling patterns and frequencies

Sampling patterns and frequencies will vary depending on the hypothesis adopted and the phase of investigation. For example during the exploratory investigation of situations covered by hypothesis H2 (e.g. a suspect leaking underground tank) sampling points should be located where the contamination is most expected to occur, and only a limited analytical suite (based on the known contents of the tank) should be employed. In the case of a pre-purchase survey, it might be possible to demonstrate the presence of the suspected contamination by judicious placing of just a few trial pits. In contrast, detailed investigation will usually require that a comprehensive picture is established of contamination over all parts of the site and the demonstration, as far as practicable, that those parts of the site believed to be uncontaminated are in fact free of contamination. Note however that proving a site is uncontaminated (i.e. satisfying hypothesis H1) is always very difficult (see Box 4.2).

Box 4.2 *Proving a site is uncontaminated*

Very often there is a requirement to satisfy hypothesis H1 i.e. that a site (or part of a site) is 'uncontaminated' or, more likely, not contaminated to an extent that 'matters' (i.e. all concentrations are below a level requiring action — see Section 6). This may be the objective of the detailed phase of investigation or the main purpose of an investigation for compliance/performance (i.e. to demonstrate that remedial action to remove or destroy contamination has been successful — see Section 7).

Proving that a site is uncontaminated is always extremely difficult and invariably requires very extensive investigation. Sampling should always be systematic (see Section 4.2.3) and testing for a wide range of contaminants may be involved. Ferguson[1,2] provides guidance on the statistical basis (including degree of confidence that specified areas of contamination will be detected) of different systematic sampling patterns.

The minimum number of sampling locations should be proportional to the area of the site. There is no justification for reducing the number of sampling locations per unit area as the size of the site increases.

The choice of sampling patterns and frequencies is also important in the investigation of the water environment. In contaminated land applications, the water environment may include surface water as well as groundwater, and the strategy will need to be adjusted to incorporate both where applicable. As a general rule, comparatively fewer water than soil samples are normally taken because a water sample, provided it is from the same water body, is generally more representative of the parent material than a soil sample. However, establishing the condition of the water environment will usually involve repeated sampling over (extended) time periods and thus permanent or semi-permanent monitoring installations will be required.

4.2.3 Sample collection

Table 4.2 summarises the main exploratory methods available for the collection of soil and groundwater samples, and their advantages and disadvantages in contaminated land applications. Other types of sampling (e.g. flora and fauna) are dealt with in Appendix 6.

Whatever method is used to obtain a sample the equipment used should:

- not contaminate the sample (e.g. oil and paint from machinery, adhesives used in borehole construction)

- not absorb contaminants or allow contaminants to escape (e.g. sample containers)

- be kept clean to avoid cross contamination between samples[3,4,5]

Note that boreholes can be formed in a number of ways; there may be more or less opportunity to obtain relatively undisturbed samples and for the potential release of contaminants into the environment[6]. The selection of installation technique should therefore be made with care.

Table 4.2 *Methods of exploration*

Methods	Advantages	Disadvantages
Surface sampling	• Ease of sampling • Allows assessment of immediate hazards	
Augering	• Allows examination of soil profile and collection of samples at pre-set depths	• Limited depths achievable • Ease of use very dependent on soil type • Can lead to cross contamination if not done with care
Trial pits and trenches	• Allow detailed examination of ground conditions • Ease of access for discrete sampling purposes • Rapid and inexpensive	• Limitation on depth of exploration • Greater exposure of media to air and greater risk of changes to contamination • Greater potential health and safety impacts • More potential disruption/ damage to site • May generate wastes for disposal • More potential for escape of contaminants to air/ water • May need to import clean material to site for backfilling (to ensure clean surface)
Boreholes	• Permit greater sampling depth • Provide access for permanent sampling/ monitoring points • Less potential for adverse effects on health and safety, or above ground environment (but note potential risks to groundwater) • Smaller volumes of waste to dispose of • May permit integrated sampling for contamination, geotechnical and gas/water sampling	• More costly and time-consuming • Less amenable to visual inspection • Limited access for discrete sampling purposes • Depending on the technique may be disturbance to samples and therefore loss of contaminants • Potential for contamination to an underlying aquifer • Potential for groundwater flow between strata within aquifer
Driven Probes	• Minimal disturbance of site-no need to remove material from the hole • Some soil properties can be determined during penetration • Undisturbed samples can be recovered • Variety of measuring devices can be installed once hole is formed • Fewer health and safety, and above-ground environmental implications	• No opportunity to inspect strata • High mobilisation costs for most powerful equipment
Soil gas sampling	• Can detect volatile organics • Allows contamination plumes to be mapped • Avoids difficulties of handling and analysing samples for volatiles	• Can detect only volatiles and semi-volatiles • Semi-quantitative data only
Groundwater sampling:		
• standpipe/well	• Permits water levels, water quality and permeability to be measured and pump tests to be carried out	• Cannot provide accurate water quality, levels or pressure data in sequences of strata of different permeability • May permit cross-flow within and between aquifers
• 'nested' wells and similar installations	• Permit sampling from different depths	• More expensive
• piezometer	• Permits monitoring of water pressures and permeability to be tested, particularly over specific zones	• Not usually suitable for groundwater quality testing

It may be necessary to break-out concrete or other hard-standing to sink boreholes, excavate trial pits/trenches or to insert probes. This should be done with care: liquid contaminants may have accumulated beneath the concrete, or contaminated materials used as hardcore or fill.

4.2.4 General sampling requirements

General requirements applying to all types of sampling of contaminated (or potentially contaminated) material include the need:

- to define the position of each sample accurately, both in plan and by depth

- to ensure sufficient samples are taken, including samples for additional testing purposes

- to collect samples of sufficient size in suitably prepared, inert, robust, securely sealed containers

- for accurate, permanent labelling of samples

- for secure and timely transport of samples to the testing laboratory, since time delays may affect the validity of some analytical results

- to comply with relevant regulations regarding the transport of hazardous materials (see Volume XII)

- for careful handling and storage of samples at all times, including maintaining an appropriate storage temperature (generally between 2 and 4°C). Insulated containers specifically designed for transport of soil and water samples should be used — recreational coolers do not give sufficient temperature control[7].

- for careful and accurate field description of collected samples

- to record conditions in the field, such as the presence of odours, representativeness of the sample, and any particular features or characteristics present which were not, or could not be, sampled

- to maintain a photographic record of general site conditions, of conditions in trial pits etc. and of the conduct of the site investigation

- to protect the health and safety of workers, public health and the environment

- for the personnel engaged in sampling to be appropriately qualified and experienced.

Requirements for investigation personnel are given in BS DD 175[8] and the Dutch standard on sampling for soil quality[9] — see Box 4.3.

4.3 SAMPLING OF SOILS AND SIMILAR MATERIALS

4.3.1 General aspects

Table 4.3 summarises the typical options for sampling, and the issues to be addressed when sampling soils and associated materials.

Note that while 'spot' samples representing a narrow depth range are normally used in the investigation of contaminated land, the Dutch NVN 5740[9] provides guidance on situations during exploratory investigations where mixed or composite samples might be appropriate.

Table 4.3 *Sampling options for soils etc.*

Options	Comments
Stages of sampling :	
Single-stage	Where timescale is restricted but rarely sufficient in contamination studies
Multi-stage	The preferred approach allowing interim assessment of data and refinement of objectives
The number of sampling locations	Depends on : • size and topography of site • likely distribution of contaminants present • degree of confidence required (see references 1 & 2)
Sampling pattern :	Systematic e.g. regular grid pattern — see Figure 4.1A (usual for detailed stage of investigation) Regular grid pattern supplemented in areas of particular interest Herringbone pattern — gives higher probability of detecting elongated area of contamination than regular grid (Figure 4.3 and references 1 & 2) Systematic unaligned sampling — gives higher probability of locating isolated area of contamination but more difficult to implement in practice Stratified random sampling (see Figures 4.1B and 4.2) — of limited value Random sampling — inappropriate for contaminated sites (Figure 4.1C) Unequal/judgemental/targeted sampling (usually for exploratory or supplementary stages of investigation) — can be valuable if based on prior information regarding probable distribution of contamination. Useful as supplement to systematic sampling
Number of samples/tests from each location	Depends on variability of strata, depth of sampling etc.
Choice of sample type	'Spot' from a discrete location and depth (preferred for most purposes) 'Composite' formed from a number of sub-samples combined in the field to represent a defined area (frequently used in agriculture but not normally appropriate for contaminated land work) 'Mixed' formed by taking sub-samples from spot samples and combining them under controlled conditions in the laboratory — a portion is retained for separate analysis if required. Should never be employed with volatile contaminants or those which might react with one another.
Maximum depth of sampling	Depends on : • types of mobility of contaminants • geology and hydrology • hazards posed by contamination • depth of disturbance expected during construction • depth to natural ground underlying any made ground

4.3.3 Sampling patterns and frequency

Guidance on sampling patterns and frequencies is provided in DD175:1988[8], in the draft ISO standards for soil sampling [10,11,12] and has been reviewed by Ferguson.[2] Box 4.4 summarises the guidance provided in DD175.

Box 4.3 *Character and role of sampling personnel (after reference 9)*

Sampling personnel should :

- be highly responsible, reliable, accurate and resolute
- be able to work independently and diligently
- have practical intelligence and organisational talent
- be in good physical condition and willing to work under extreme weather conditions
- have appropriate technical knowledge and skills
- have knowledge of commonly applied tools and techniques and how they should be applied to avoid such problems as cross contamination
- be party to procurement of sampling equipment
- have detailed knowledge of health and safety requirements
- be aware of relevant environmental protection requirements
- be familiar with relevant standards and codes of practice etc.
- should know why samples are being taken and the objectives of the investigation
- should be party to the design of the sampling strategy
- should have clearly defined responsibilities
- be capable of making and recording accurate on-site records
- should be capable of carrying out necessary on-site testing/measurements.

Box 4.4 *DD175:1988 Recommendations for sampling*

DD175:1988 recommends that in general systematic sampling should be employed (e.g. taking samples from locations on a regular grid — see Figure 4.1a, or employing stratified random sampling, see Figure 4.1b and Figure 4.2).

The document makes detailed recommendations on the minimum number of sampling locations required for sites of different sizes (see below) assuming a regular grid pattern is employed. It also recommends that the size of the grid is reduced, or samples taken from selected locations based on site history ('judgemental sampling'), in areas where additional information is required. The recommendations in DD175:1988 are based on theoretical and practical considerations, and the results of field studies[13].

According to DD 175 the spacing between sample locations should depend on the size and shape of the site, and the orientation of the grid. An indication of the spacings that would result on a square site is given below. In practice an increase in spacing up to about 25 m occurs as the size of the site increases. Spacings greater than this on sites larger than 5 hectares are unlikely to provide a sufficiently detailed picture of contamination on which to base the final remediation strategy, although they may be sufficient for the first stage of a more comprehensive investigation or to demonstrate the presence of contamination to a potential vendor or purchaser.

Area of Site (hectares)	Minimum number of sampling Points	Side* (square site) (metres)	Grid Size* (metres)
0.5	15	71	14-18
1.0	25	100	17-20
5.0	85	224	22-25

Authors' interpretation

Ferguson[2] suggests *'an efficient sampling pattern should satisfy four conditions:*

(i) it should be stratified (i.e. the area to be sampled should be partitioned into regular sub-areas)

(ii) each stratum (sub-area) should carry only one sampling point

(iii) it should be systematic

(iv) sampling points should not be aligned.

The simple random pattern [Figure 4.1C] satisfies only condition (iv). The square grid [Figure 4.1A] satisfies all but one condition (ii) and stratified random patterns [Figures 4.1B & 4.2] satisfy all but condition (iii). The disadvantage of the square grid is its much reduced ability to detect elongate hot spots whose long axis is parallel to the grid axis. The weakness of the stratified random sampling patten is its tendency for uneven sampling. The herringbone sampling pattern [Figure 4.3] overcomes these disadvantages and yet retains the practical advantages of being easy to set out on site. This pattern satisfies all four conditions'.

This last ('herringbone') pattern does not appear to have been tested in the field. However it can be formed from four overlapping square grids of the same size and orientation and should not be too difficult to lay out.

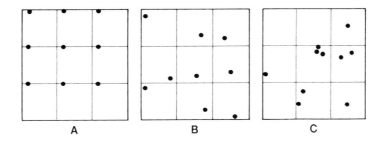

Figure 4.1 *Three sampling designs (A) Regular (square) grid (B) Stratified random (C) Simple random (after reference 2)*

Ferguson addresses the number of sampling points needed to detect a 'hot spot' of contamination (defined as a % of the total area of the site) with a specified degree of confidence (e.g. 95%) such that :

- a hot spot (if one exists) will not be missed

- if contamination is not found, a hot spot of at least the specified size does not exist.

The size of hot spot to be targeted should be based on consideration of the largest hot spot that could be dealt with economically (and without unacceptable health risks) were it to be missed during the site assessment[2] (e.g. were it to be exposed during remediation works). It is important to bear in mind the area of contamination which might be of concern in the estimation and evaluation of human health risks (see Section 5). For example, the primary risk is often considered to be to small children ingesting soil while playing in their own gardens: it might therefore be considered appropriate to devise a sampling strategy that will detect an area of contamination equivalent to a small garden (which may be only 50 m^2 or just 0.5% of a 1 hectare site — see Box 4.5). Figure 4.4 shows the relationship between a grid pattern of sampling and a typical modern housing development. Figure 4.5 shows the ability of different sampling designs to detect a circular target occupying 5% of the total site area. Figure 4.6

shows the comparative performance of square grids and herringbone patterns for different target shapes: the regular grid might be judged satisfactory (provided sufficient samples are taken) unless there is an elongated target aligned with the grid.

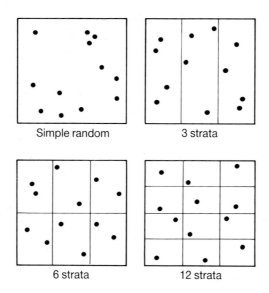

Figure 4.2 *Examples of stratified random sampling (16 Sampling locations)*

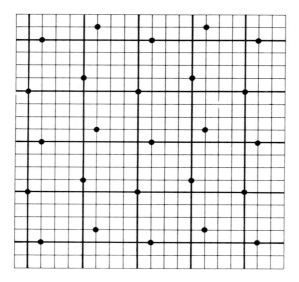

Figure 4.3 *Herringbone sampling design*

In practice, the probability of detecting a hot spot of defined size can be improved by use of multi-stage sampling[2].

The size of the 'hot spot' is not a fixed parameter and is to some extent in the control of the investigator. The size of the target area may vary therefore depending on the definition of a hot spot. A hot spot may be regarded as:

- an area of contamination within an otherwise uncontaminated site
- an area of greater contamination within a site which is generally contaminated.

Figure 4.4 *Relationship between a regular sampling grid and a typical modern housing development*

Box 4.5 *Identifying an area of contamination with 95% confidence*

To locate (with 95% confidence) a circular target area of contamination of 100 m^2 (i.e. equivalent of two small gardens) on a 1 hectare site employing a herringbone sampling pattern would require about 110 sampling locations if a single-stage sampling programme were employed (see Figure 4.7). This contrasts with the 25 sample locations suggested in DD175:1988.

However, 25 sample locations on either a regular square grid, or a herringbone pattern, would be sufficient to locate a circular target of 500 m^2 (5% of site) with 95% confidence.

If the target area is elliptical (e.g. aspect ratio 1:4) the probability of locating a 500 m^2 target using a regular square grid could fall to less than 70% if the target was aligned parallel to the grid (see Figure 4.8).

For a 100 m^2 elliptical target and 25 sampling points, the probability of locating the target would be less than 30% for both sampling patterns.

The comments in the text regarding the concept of target/hot spot size should be noted.

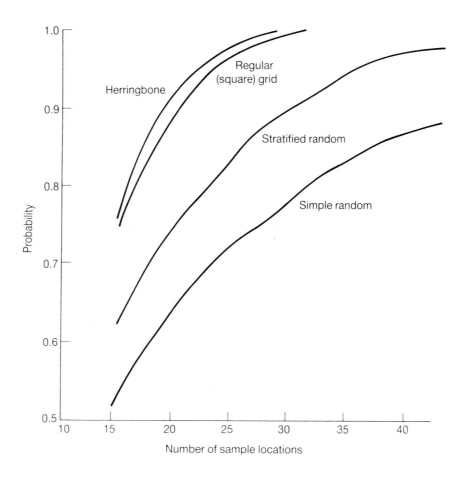

Figure 4.5 *Performance of four sampling designs for detecting a circular target occupying 5% of total site area (after reference 2)*

It might also be defined, for example, as an area with contamination:

- above a guideline 'trigger' concentration
- say, two standard deviations above 'background'

- in excess of a guideline mandating remediation
- above some arbitrary value.

It is important to bear in mind that the sampling strategies developed by Ferguson apply only to a single contaminant in a single plane. The distributions of different contaminants on a site may vary because they have different origins and, if from the same source, because they behave differently in the ground. Appropriate strategies for sampling at depth must also be developed.

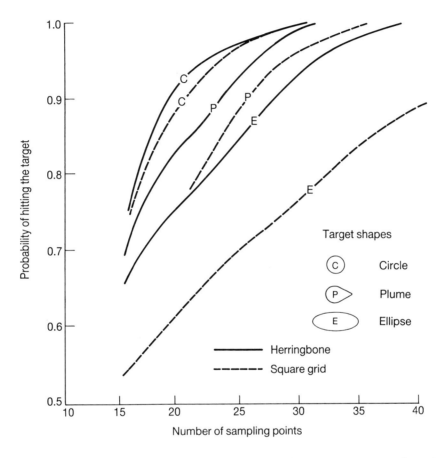

Figure 4.6 *Performance of square grid and herringbone sampling designs for detecting targets of various shapes. Relative size of each target is 5% of total site area*

Many contaminated sites are extremely heterogeneous and variations in concentrations of individual contaminants may show little systematic pattern and can be markedly different even over distances of a few centimetres. Figures 4.9 to 4.11 show the results obtained in one experimental study[13].

4.3.3 Sampling procedures

Sampling procedures for contaminated soils etc. depend on the exploratory technique(s) used. Important common requirements include :

- Sample size should be a minimum of 1 kg. Larger samples may be required for special tests (e.g. detection and characterisation of metallurgical slags, see Appendix 6)

- Disturbed samples (for chemical and similar analyses) should be collected separately from those required for other purposes (e.g. geotechnical testing)

- Sample containers should be sealable and made of materials that will not react with or absorb contaminants (see Table 4.4)

- All containers should be thoroughly cleaned before use

- Sample containers should be filled to capacity to minimise the volume of air, unless the sample has the ability to produce gas (in this case a safety pressure valve should be fitted to reduce the risk of explosion during transport)

- Where the presence of volatile substance is anticipated, particular care is required in the collection and handling of samples : undisturbed samples should be taken wherever possible (e.g. using a sealable coring device that is only opened under laboratory conditions; the laboratory should be consulted about the type of container to be used).

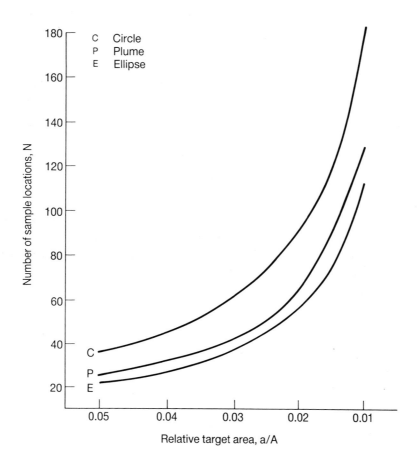

Figure 4.7 *Number of sampling locations needed to ensure 0.95 probability of success in hitting targets of different relative sizes (after reference 2)*

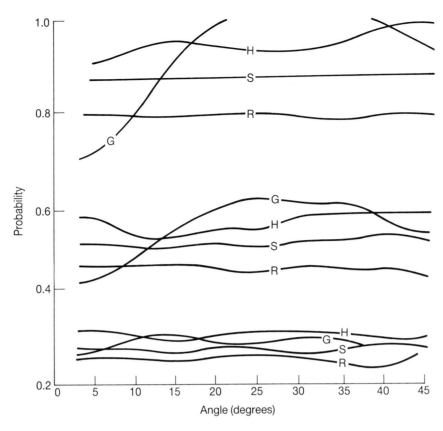

Key: H = Herringbone G = Regular (square) grid
     S = Stratified random R = Simple random

Figure 4.8 *Performance of four designs for detecting an elliptical target (aspect ratio 4:1) as a function of orientation. Relative size of target is 5% (top) 2% (middle) and 1% (bottom) of total site area. Number of sampling locations is 30 (after reference 2)*

Table 4.4 *Types of sample container*

Types of container	Applicability
• glass	All purpose
• stainless steel	
• aluminium	
• rigid polyethylene	If organic contaminants are unlikely to be present
• polypropylene	
• polythene bags	Not recommended on grounds of:
	— difficulty of achieving an effective seal
	— potential permeability to some contaminants
	— possibility of damage during transport

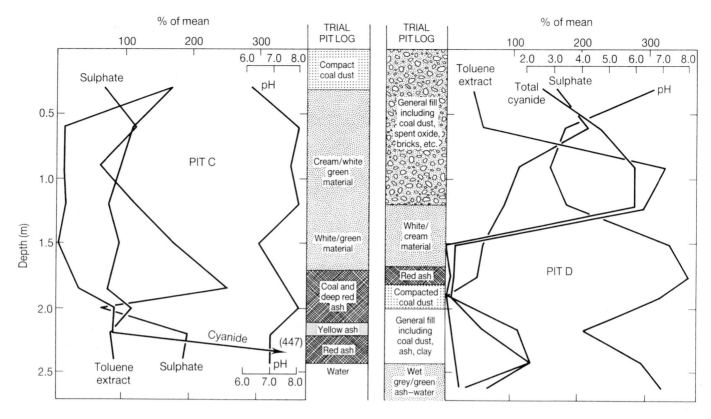

Figure 4.9 *Variation of contamination with depth and relationship to materials sampled for trial pits C and D (reference 13)*

Research has shown that even under ideal laboratory conditions no more than about 50% of volatile substances may be retained. Under worse case conditions (e.g. disturbed samples placed in polythene bags) virtually all the organics may be lost (see Table 4.5)[14]. Specialist guidance should be sought for these substances. Explosives residues should be sampled only by appropriately qualified personnel[15]. Appendix 6 contains more detailed information on sampling specific material types.

Key to Table 4.5 (opposite):

A = Undisturbed sample (core) in Teflon-sealed glass jar with headspace equal to 85% of total volume of jar.

B = Undisturbed sample in Teflon-sealed jar with headspace volume equal to 40% of jar volume.

C = Undisturbed sample immersed in methanol (about 40 total volume) in Teflon-sealed jar.

D = Disturbed sample (taken in 7-10 aliquots with stainless steel spoon in Teflon-sealed glass jar with headspace volume equal to 40% of jar volume.

E = Disturbed sample (taken in 7-10 aliquots with stainless steel spoon) in empty laboratory-grade plastic bag with zip closure and headspace of about 40% of volume

* = Best estimate of concentrations in laboratory prepared material that was sampled.

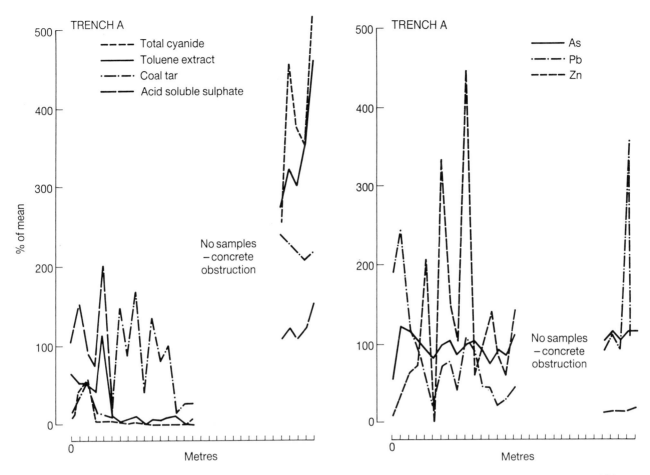

Figure 4.10 *Variation of contamination along length of Trench A (reference 13)*

Table 4.5 *Results of experimental studies on sampling soils for volatile organics*[14]

Contaminant/sampling method/sample treatment						
	E	A	D	B	C	Original*
Disturbance	yes	no	yes	no	yes	
Headspace	low	high	low	low	low	
Container	bag	glass	glass	glass	glass	
Methanol	no	no	no	no	yes	
Measured level of contaminant *(All values mg/kg)*						
Methylene chloride (MC)	<0.4	1.75	6.10	4.90	7.2	24.5
1,2 dichloroethane (DCA)	<0.1	5.15	5.15	6.70	18.72	22.6
1,1,1-trichloroethane (TCA)	<0.1	0.20	0.28	0.36	1.87	6.6
Trichloroethylene (TCE)	0.1	0.32	0.42	0.55	2.27	4.7
Toluene	0.06	0.37	0.39	0.49	0.70	1.7
Chlorobenzene	<0.01	0.56	0.58	0.69	0.76	1.5

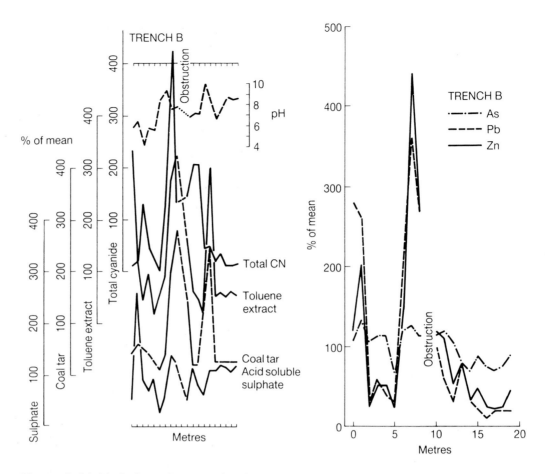

Figure 4.11 *Variation of contamination along length of Trench B (reference 13)*

4.4 SAMPLING GROUNDWATER

4.4.1 Sampling strategies

The placing of sampling/monitoring installations must take account of the extent of contamination of the groundwater and the location of known discharge points, where these exist. Information on such discharges may be available from the preliminary site investigation or be predicted by modelling techniques.

Because hydrogeological characteristics, and the nature and likely distribution of contamination, are all relevant to the design of monitoring wells, a preliminary investigation and sampling programme will be necessary before designing any long-term monitoring network. It may not be possible to finalise the detailed design of the monitoring wells (e.g. location and length of screened section(s) etc.) until drilling is complete. There will often be a need, therefore, for a degree of flexibility in technical specifications and contract details. In complex strata, or where the nature and predicted behaviour of the contamination requires access to different sampling depths (e.g. stratified systems), a number of installations of varying depth (or nested installations) may be required.

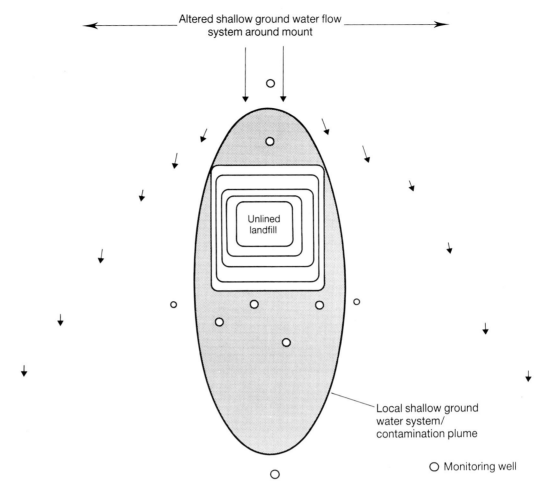

Figure 4.12 *Mounding of water within a landfill causes modification of local groundwater flow patterns (adapted from reference 19)*

Installations should be located both inside and, where possible, outside the site since this can assist in characterising hydrological conditions generally. Monitoring points should be located both upstream (to provide control data), and downstream of the source since most dissolved constituents will move vertically downwards through the unsaturated zone beneath the source area, then horizontally in the direction of flow when encountering the saturated zone. The groundwater table may be elevated locally within a landfill or close to a leaking sewer so that local groundwater flow directions may differ from regional patterns (see Figure 4.12). DNAPLs (Dense Non-Aqueous Phase Liquids) can also move against the groundwater flow direction (see Figure 4.13). Downstream installations should be sited in the most permeable strata underlying the source area.

The minimum number of sampling locations should be proportional to the area under investigation. There is no justification for reducing the number of sampling locations per unit area as the size of the site increases.

Phasing can provide an economic approach to groundwater investigation (see Table 4.6).

Table 4.6 *Phasing groundwater investigations*

Phase of investigation	Sampling/monitoring activities
Exploratory investigation	• construction of limited number of installations within and around the site based on preliminary investigation data[a] • measurement of water levels • preliminary water quality measurements
First phase of detailed investigation	• construction of additional monitoring installations on grid basis to give broad cover across area of interest[b] • further monitoring of levels • water quality measurements
Detailed (and subsequent) investigations	• further adjustment of monitoring network where appropriate based on findings[b] • in-situ testing (e.g. pump testing, permeability measurements etc. to determine aquifer properties) • water level measurement • groundwater quality measurements

Notes :

a. Installations used may be piezometers (e.g. to determine water levels/pressures) or simple standpipes (for preliminary water quality measurements) depending on objectives

b. Earlier findings should be used to determine location, depths and types of installation required.

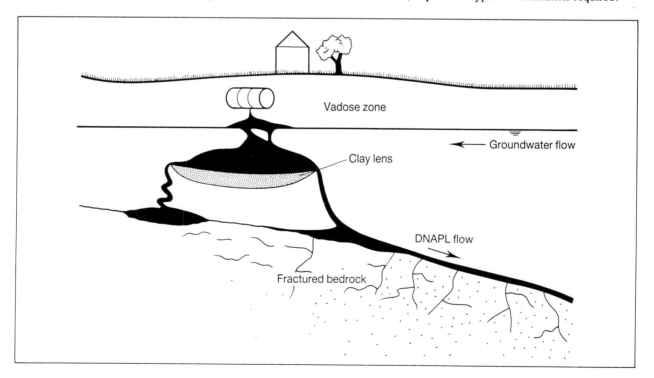

Figure 4.13 *Subsurface distribution of dense non-aqueous phase liquid (DNAPL)*

4.4.2 Types of installation

The type of installation used depends on the objectives of the work. It is essential to define both short- and long-terms objectives (e.g. preliminary sampling only, long-term monitoring, access for in-situ testing etc.) at the outset if the right type and quality of data are to be obtained[16].

Materials and methods of construction should

- have minimal effect on water chemistry
- be sufficiently robust to withstand the potentially harsh (physically and chemically) environment found on contaminated sites.

Table 4.7 summarises information on the types of installation that may be used in groundwater investigations. More detailed information and guidance on construction methods and materials is given in Appendix 6.

Table 4.7 *Types of groundwater monitoring installations*

Type of installation	Applications
Simple standpipes (various diameters, section lengths etc.)	• measurement of water levels • preliminary water quality when installed to correct depth • not appropriate for measurement of water heads, particularly in sequences of strata of alternating permeability where water is contained under different pressures including, in some cases, under artesian conditions
Large diameter standpipes or observation wells (diameter up to 150 mm, 50-100 mm sizes in common use)	• monitoring of water levels • sampling for water quality, including over long-term • permeability testing • pumping tests • larger types could be used subsequently for remedial purposes (e.g. pump-to-treat) or where investigation to very considerable depths is required
Piezometers (small diameter, typically 19 mm casing, with porous tip for measuring pore pressures; different types available including various response time, manual/automatic recording etc.)	• essential for accurate monitoring of water pressures, particularly over specific zones • measurement of permeability in specific zones

Detailed guidance on the various types of installations used in groundwater monitoring applications is given in a number of standard texts [4,5 and 18-23] (see also documents listed in Appendix 1).

4.4.4 Sampling techniques

Sampling involves collection of data on both physical and chemical properties. It is usually a three phase process :

- determination of water levels
- purging
- sampling.

The sampling strategy must therefore address:

- any pre-purging of the installation to ensure the collected sample is representative of the groundwater (see Box 4.6)

- procedures used to collect and contain the sample pending analysis (see Box 4.7).

Box 4.6 *Purging*

The well may have to be purged before sampling so that samples are representative of the 'true' groundwater rather than water that has been stored in the well (and hence subject to potential reactions with air or borehole lining materials (e.g. desorption and absorption). The amount and rate of purging depends on the hydraulic characteristics of the surrounding geology, well construction details and the sampling methods used – purge volumes and pumping rates should therefore be evaluated for each specific site. Purging should be sufficient to ensure the representativeness of the sample, while minimising disturbance of the regional flow system or the collected sample (note that turbulence may result in the loss of volatile components). Sampling without purging may be adequate to determine suitable disposal arrangements for the purged waters.

Box 4.7 *Sampling groundwater*

- Sampling should commence only when conditions in the well have stabilised as indicated, for example, by pH measurement, electrical conductivity etc.
- Care should be taken in the selection of materials used in pump components and other sample delivery equipment (tubing, bailers etc.) to minimise contamination/modification of the sample
- Sampling procedures should minimise the potential for cross contamination: e.g. sampling upstream before downstream locations with the order of sampling in the latter dictated by actual or suspected contaminant types and concentrations
- Sufficient sample should be obtained for laboratory analysis – a minimum of 1 litre is usually required
- Test-specific samples should be collected in a systematic manner with samples for volatiles, total organic carbon (TOC), and determinands requiring field filtration or determination, collected prior to large volume samples for macroelements, extractable organic compounds and metals etc.
- Some analytical determinations are best made on-site, e.g. pH, conductivity, temperature, dissolved oxygen, ammonia (note that in principle field testing should be subject to comparable quality control/quality assurance procedures to those used in the laboratory although in practice it may be necessary to accept some degree of compromise in field situations)
- Samples should be 'preserved' on-site as necessary
- Sample containers should be of appropriate materials and prepared for use
- Post-sampling storage and handling procedures should be appropriate to the determinand under consideration
- Potentially hazardous samples should be appropriately labelled and protected from physical damage during transport
- Analysis should be undertaken promptly after collection of the samples, and in any event within 24 hours, to prevent deterioration before testing
- Appearance, odours and location in the field should be included in the field description of the sample
- Care should be taken in the collection of non-water-miscible liquids.

When selecting equipment a distinction should be made between sampling and purging. A variety of devices may be used for purging, but only a few can provide water samples for accurate analysis. In purging the aim is to remove water efficiently; in sampling the primary objective is accuracy with low flow rates preferred so that sample agitation and aeration are minimised.

Stainless steel and PTFE are preferred construction materials for sampling devices, although PVC construction materials are acceptable in many applications. Connecting tubing should preferably be manufactured of, or lined with, PTFE. Either dedicated or portable sampling/purging devices may be used; non-dedicated sampling devices should be thoroughly cleaned between sampling events. The performance of a selection of sampling devices is summarised in Tables 4.8 and 4.9.

Further guidance on sampling and purging is given in references 4 and 5 and 16 to 20. Guidance on the handling, storage and preservation of samples is given in BS 6068: Section 6.3[24].

Table 4.8 *Performance of sampling devices (after reference 25)*

Device	Accuracy/ Precision	Flow rate	Lift capacity (m)	Available in required materials?	Applicability to sampling
Grab samplers (thieves, bailers etc.)	Fair	Low	Theoretically none	Yes	May affect parameters sensitive to dissolved gas composition, e.g. pH, alkalinity, redox dependent trace metals
Suction lift	Low	High	Typically 5-6	Yes	Not recommended for gas sensitive parameters
Air displacement	Fair	Low	Up to 75	Yes	Accuracy reduced for gas sensitive parameters
Electric submersible	Fair	High	Up to 600	Yes	Accuracy fair for gas sensitive parameters
Piston pumps	Fair	Low	Up to 185	Yes	Accuracy and precision dependent on operator
Bladder pumps	Good	Low	Up to 300	Yes	Highest accuracy and precision

Table 4.9 *Performance of sampling devices in relation to contaminant types (after reference 26)*

Device	VOCs, organo-metallics (e.g. chloroform, TOX, CH_3Hg)	Dissolved gases Well purging parameters (e.g. pH, Eh)	Trace inorganic metal species (e.g. Fe, Cu) Reduced species (e.g. NO^{2-}, S^{2-})	Major cations and anions (e.g. Na^+, Ca^{2+}, Cl^-, SO_4^{2-})
Grab samplers (thief or dual check valve bailers)	May be adequate if well purging is assured	May be adequate if well purging is assured	May be adequate if well purging is assured	Adequate for cations May be adequate if well purging is assured
Suction lift	Not recommended	Not recommended	May be adequate if materials appropriate	Adequate
Gas drive devices	Not recommended	Not recommended	May be adequate	Adequate
Mechanical positive displacement pumps	May be adequate if design and operation controlled	May be adequate if design and operation controlled	Adequate	Adequate
Positive displacement bladder pumps	Superior performance for most applications	Superior performance for most applications	Superior performance for most applications	Superior performance for most applications

4.5 SAMPLING SURFACE WATER

Samples may be collected from a variety of moving and 'static' surface waters including rivers, estuaries, canals, ponds, lakes, disposal lagoons and docks. The aim will usually be to establish concentrations or loads of specified physical, chemical, biological or radiological parameters at selected locations throughout the whole or part of a water course or body of water.

Table 4.10 *Possible approaches to sampling surface waters*

Method/activity	Applicability/comments
Spot sampling	A single sample taken at a fixed location and depth using hand equipment (e.g. baler). Time variable samples may be obtained using automatic equipment. Recommended to detect pollution when unstable parameters are present e.g. where flow is not uniform, parameters are variable and composite samples would obscure variations
Periodic samples (discontinuous):	
• at fixed time intervals	Provides time profile of contamination
• at fixed flow intervals (volume dependent)	Taken when quality is not related to flow rates
• at fixed flow intervals (flow dependent)	Taken when quality is not related to flow rates
Continuous samples:	
• at fixed flow rates	Contain all constituents present during sampling period but do not provide information about variations in concentrations of individual parameters
• at variable flow rates	The most precise method of sampling flowing water if both the flow rate and contaminant concentrations vary significantly
Depth integrated sampling	Provides a composite water sample representative of the vertical profile; obtained by lowering and then retrieving a sample bottle at constant velocity
Point sampling at selected depths	Obtained using sample bottles opened at required depths and then retrieved (evacuated or air-filled bottles, flow through samplers or automatic samplers with inlets at specified depths may be used)
Depth profile samples	Series of samples from various depths at a single location
Area profile samples	Series of samples from a particular depth at various locations
Grabs or dredgers	Used to obtain samples of sediment — designed to penetrate under own weight
Cores	Provide information on vertical profile of sediment

The sampling programme must take into account:

- possible variations with location, depth and time of sampling

- variations in flow and level arising for example from tidal movements, seasonal fluctuations in groundwater levels, and short term fluctuations associated with rainfall events, and upstream discharges (these should be the subject of separate study)

- other factors such as temperature, seasonal fauna and flora variations (e.g. algal blooms) and the movement of boats etc.

A series of British Standards[21-24, 27-29] give guidance on sampling procedures for water quality (see also Appendix 1).

Possible approaches to sampling are summarised in Table 4.10 (see also Figure 4.14). Before sampling it is essential to consult the analyst to establish suitable sample preservation and handling arrangements[24]. The types of data required (e.g. concentrations, loads, maximum and minimum values over time or space, arithmetic means, median values etc.) must also be decided since these determine both the location and method of sampling (e.g. spot versus time composite samples). Guidance is available on the statistical aspects of sampling in relation to time and frequency[21].

To completely characterise the hydrological regime, and to establish the likely consequences of the presence of a contaminated source on users, habitats and interconnected water systems, additional data must be collected, for example:

- an estimate (where practicable) of the flow and proportion of surface or rain water in the sampled water body

- as appropriate, a description of the depth and nature of the geological stratum (in contact with the water body) from which the sample was collected

- rainfall measurements

- the physical properties of surface water bodies.

Several established procedures are available for the measurement of physical characteristics (e.g. flow using simple plate weirs, direct gauging etc.) of surface water bodies, including springs and streams etc. Larger bodies are more likely to be already gauged and the required information readily available. Meteorological data (precipitation rates etc.) are generally available from the Meteorological Office or local authority sources, or from on-site measurements.

Chemical testing parameters will in general be the same as for groundwaters although a separate and independent measurement of (suspended) solids content may be required. Note that suspended solids content typically varies markedly across water courses and depth integrated samples should therefore be taken at several locations. Preliminary on-site measurements of a limited range of parameters, e.g. temperature, electrical conductivity and pH, at numerous locations can be used to establish the homogeneity of a surface water body, and can indicate appropriate sampling locations for detailed analysis.

Sampling of water courses should:

- be at locations where there is good mixing but not excessive turbulence as this can lead to loss of volatiles, dissolved gases and oxidation of some compounds

- be in locations with sufficient flow rates to avoid stratification due to temperature or density differences (e.g. not close to banks unless of special interest)

- not take place at the immediate surface as this may be unrepresentative of the water body as a whole (subsurface samples from within 0.5 m of surface generally preferred)

- preferably be at flow measuring points so that flow and concentration data can be combined to provide an estimate of contaminant flux.

Sampling programmes for estuarine waters should take into account tidal currents and the way they are influenced by wind, density, bottom roughness, closeness to the shoreline, and the movement of shipping, discharges etc.

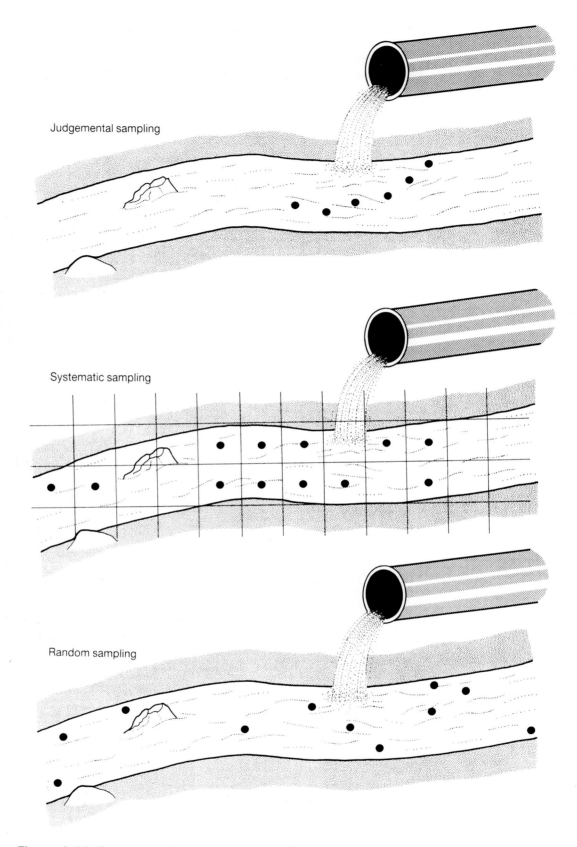

Figure 4.14 *Examples of judgemental sampling (top), systematic sampling (centre), and random sampling approaches (bottom)*

Sampling static water bodies may present difficulties due to:

- difficult access
- size and shape (e.g. an irregular shape, such as a bay or inlet, may contain 'stagnant' waters)
- stratification
- safety (depth of water, hazardous materials on surface or in sediments which may be disturbed).

Where there is free surface product, a surface film or obvious layer of suspended matter it must be decided whether to obtain samples of the layer itself or samples representative of the water body as a whole; obtaining a representative sample is usually only possible where the water is well mixed (e.g. over a weir), but note the potential for loss of organic compounds at such locations.

Sediments are typically layered. Physical properties (e.g. particle size distribution), chemical composition and chemical conditions (e.g. pH and redox potential) tend to vary with both location and depth. This variation can only be properly addressed by taking a large number of samples. Hand tools can be used in shallow depths of water to obtain samples of surface sediments. For deeper or fast flowing waters, or to obtain sediment profiles, grabs or coring devices must be used (see Table 4.10).

4.6 ANALYTICAL AND TESTING STRATEGIES

Analytical and testing strategies should take account of:

- the range of tests or analyses to be carried out, the methods to be employed and detection limits required
- the selection of samples to be analysed from among those collected from the field
- whether testing or analysis is to be performed in-situ, on-site or off-site.

Clearly it is not feasible to analyse a large number of samples for all possible contaminants and some means of rationalising the analytical strategy is usually required. The strategy should be based on :

- the objectives of the current phase and of the investigation as a whole (e.g. to detect contamination, determine physical properties that may affect the choice of remedial method)
- previous investigation findings
- observations made on site
- visual inspection of samples.

Table 4.11 summarises the factors that should be considered when developing analytical/testing strategies.

Depending on site-specific circumstances, a comprehensive approach to chemical analysis is likely to involve:

- testing for ubiquitous contaminants, such as lead, zinc, and mineral oils
- testing for substances that are specific to the past, and present, use of the site as indicated by the desk study
- the use of screening techniques to detect adventitious substances and those whose presence cannot easily be predicted.

Table 4.11 *Analytical design parameters*

Parameter	Variable
Scope of the analytical programme	Range of tests Numbers of samples Types of samples
Use of screening techniques	Field based Laboratory based
Sample preparation needs	As received, air dried, other Size reduction Size fraction Extraction
Detection method	Sensitivity Reproducibility Turn-round time Reliability Cost
QA/QC procedures	Sample logging Blank samples Spiked samples Recovery/ accuracy performance Storage of samples Disposal of samples

---

Box 4.8 *Environmental analysis*

Environmental analysis comprises two separate disciplines of organic and inorganic analyses involving quite different approaches to answering the questions:

- What is present ?
- How much is present in a particular sample ?

In the inorganic laboratory, the analyst is primarily concerned with the analysis of a defined number of elements (metals and metalloids) and anions. In contrast, the organic analyst may be asked to carry out analysis for almost any organic chemical which might happen to be present in the soil or water sample. It may be necessary to screen the sample for the presence of forty, fifty or even hundreds of specific compounds.

Inorganic analytical methods are generally more clearly defined than those for organic contaminants. Automation and computerised data handling have increased the availability of relatively rapid, sensitive and low cost analyses. In contrast, organic analysis is slower and more expensive; individual samples typically take several days at least to complete and costs per determination may be higher by an order of magnitude.

In practice organic analysis takes two main forms:

- qualitative analyses which attempt to answer the question 'what is in the sample ?'
- quantitative which attempts to answer the question, 'how much of a specific compound or class of compounds is present in the sample ?'

Some methods seek to identify and quantify classes of compounds (e.g. chlorinated hydrocarbons, petroleum hydrocarbons) rather than individual compounds.

Qualitative analysis is frequently carried out prior to quantitative analysis. Class analyses frequently precede specific compound analyses.

In principle, and at least initially, it is better to undertake a smaller range of key tests on a large number of samples, than elaborate and expensive testing on only a few samples. A staged approach can also be effective and economic. For example, 'total' concentrations may be used as initial indicators of the presence of particular contaminants, with analysis for specific, and perhaps more reactive or mobile, chemical forms triggered only when specified total concentrations are exceeded. Water-soluble sulphate, free cyanide and monohydric phenols are all examples of determinands commonly tested as 'dependent options' in this way. However, decisions to use such 'triggers' (and the concentrations at which they apply) should be made separately for each site in the light of what is known about the probable nature of the contamination. It is inappropriate to include such triggers in standard specifications.

The retention of samples to be tested on the basis of the findings of the initial analysis is a further option for consideration, provided the contaminants and materials of concern are not adversely affected by storage.

The detection of adventitious or unexpected substances, particularly when complex mixtures of organic chemical species are present, requires the use of analytical screening methods. Techniques such as gas chromatography/mass spectrometry (GC/MS) and inductively coupled plasma spectrometry (ICPS), can provide data on a wide range of compounds and elements and be a useful means of identifying more specific analytical requirements (see Box 4.8).

Other factors that determine the detailed scope of the analytical or testing strategy include :

- the use of specific extraction regimes to provide measures of 'total', biologically 'available', leachable or water-soluble contaminant concentrations (see also reference 30)

- sample preparation needs specific to different types of contaminants

- the accuracy and detection limits of the analytical methods used

- quality management procedures employed to ensure the validity of the result (see Section 3).

Analytical laboratories frequently offer analytical packages (see Box 4.9) covering a range of common contaminants: these (and in particular any 'conditional' provisions or the application of surrogate or broad spectrum analyses) should be carefully checked to confirm they are consistent with site-specific requirements as determined by the desk study or previous investigation findings.

Box 4.9 *Typical analytical package for soils*

Elements	Anions	Other determinands
cadmium	chloride	pH
lead	sulphate	phenols
arsenic	sulphide	toluene extractables
chromium		cyclohexane extractables
zinc		coal tars
copper		mineral oils
nickel		sulphur
boron		polyaromatic hydrocarbons (PAHs)
mercury		electrical conductivity
		BTEX (benzene, toluene, ethylbenzene, xylene)
		total hydrocarbons

The analysis carried out must clearly cover any statutory requirements. Experience and consideration of site histories should be used to predict the principal contaminants likely to occur on particular types of site (see Box 4.10 for a list of contaminants commonly associated with coal carbonisation sites). The fact that a potential contaminant is not covered by generic assessment criteria (e.g. ICRCL guidelines) is not an acceptable reason for failing to test for a potentially hazardous substance.

Box 4.10 *Typical gasworks contaminants*

- coal tars constituents – PAHs, benzene etc.
- phenols
- cyanides – simple and complex
- sulphur and its compounds – sulphates, sulphides, thiocyanates etc.
- ammonia and its compounds
- mineral oils
- coal/coke residues
- acidity
- toxic elements – arsenic, copper, zinc, mercury etc.

It is possible to develop matrices of 'typical contaminants' associated with 'typical land uses'. Information on contaminant/land use relationships is available from a number of sources (see Appendix 2). However, sites may have been used for a number of purposes and a review of the site history, and activities on neighbouring land, should always be carried out to determine what other contaminants might also be present.

'Generic' extraction techniques for 'organic' contaminants should be used with care. for example, toluene extractable matter (TEM):

- may be elevated due to the presence of a wide range of organic compounds: not only the 'mineral oil' and 'coal tar' that is frequently determined as dependent options

- will not detect the presence of toluene or other volatile compounds

- may be elevated due to the extraction of 'humic' substances from peat or other natural 'humic' constituents

- unless corrective measures are taken during the analysis also include elemental sulphur and organosulphur compounds.

Dependent option testing based on the TEM, for example being above 2000 mg/kg, will be seriously misleading if the organic substances extracted include compounds for which the concentrations of concern are lower than this threshold point: for example 50 mg/kg for PAHs (ICRCL threshold trigger value), PCBs say 1 mg/kg. In general, the TEM method commonly employed is not capable of reliably detecting concentrations below a few hundred mg/kg. It is, therefore, preferable to use more specific and sensitive screening methods.

4.6.1 On-site measurements

Rapid on-site methods of chemical analysis have been reviewed by Montgomery *et al*[32]. The issues covered include:

- quality assurance
- compatibility testing
- air monitoring using instrumental techniques and detection tubes
- toxicity testing
- specialised screening techniques for soils (e.g. for PCBs, phenols)
- equipment requirements for field laboratories.

On-site analysis of hazardous wastes is also addressed by Simmons[33] and specialist conference proceedings.[34,35]

4.6.2 Off-site laboratories

Off-site analysis may be carried out by the investigating organisation or by another body under contract to the investigating organisation. In either case, there should be early liaison with the laboratory to discuss sampling and analytical arrangements because inappropriate sampling methods, sample handling and pre-analysis sample preparation can severely limit the value of subsequent analytical effort.

Any laboratory carrying out contamination related analyses should be comprehensively equipped, although few will be able to offer all the specialist techniques that might be required. Organisations offering laboratory services should be able to demonstrate that they:

- have experience of the methods to be applied
- understand the context of the work to be carried out
- are able to employ methods that are sufficiently sensitive (in terms of minimum detection limits) and analytically reliable (in terms of reproducibility and accuracy) for the task in hand
- are familiar with standards for sample preparation and analysis and are able to apply them
- have appropriate quality assurance and quality control measures in place
- are accredited under NAMAS and other appropriate schemes
- participate in appropriate inter-laboratory proficiency schemes (see Section 3)
- are able to meet any protocols for sampling and sample handling necessary to assure preservation of the sample and safety of all who may come into contact with samples.

Contract requirements should be explicit (e.g. a simple test result only, or interpretation of the data). Except where the latter is very limited in scope (e.g. whether some control limit is exceeded) an additional fee is likely to be charged for what amounts to advice and consultancy.

4.6.3 On-site laboratories

On-site laboratories are most likely to be used when an investigation involves:

- only a limited number of analytical determinands selected on the basis of site history and earlier investigation

- rapid analysis of indicator species as a means of controlling on-site works (e.g. extent of investigation, remediation) with subsequent confirmatory testing in an off-site laboratory.

On-site laboratory facilities are likely to be essential for remedial strategies involving on-site process-based treatment of soil or groundwater to:

- control the feed to the process
- monitor the 'product' from the process
- monitor effluents and emissions.

Field testing and on-site laboratories can reduce turn-round times and provide data that can be used to direct site works (e.g. extend the depth of a borehole, length of a trench). However, analytical reliability may be reduced, and there may be a need to employ surrogate methods that have been shown to provide an adequate indication of concentrations of key contaminants.

REFERENCES

1. FERGUSON, C. A statistical basis for spatial sampling of contaminated land. *Ground Engineering*, 1992, June, 34-38

2. FERGUSON, C. *Sampling strategies for contaminated land*. CLR Report No. 4 Department of the Environment, (London), 1994

3. STEEDS, J., SHEPHERD, E. and BARRY, D.L. *A guide to safe working practices for contaminated sites*. Report 132. CIRIA (London), (in the press)

4. UNITED STATES ENVIRONMENTAL PROTECTION AGENCY. *Handbook of suggested practices for the design and installation of groundwater monitoring wells*. USEPA (Washington DC), 1991, EPA/600/4-89/034

5. AMERICAN SOCIETY FOR TESTING AND MATERIALS. *Standard practice for the design and installation of groundwater monitoring wells in aquifers*. ASTM D5092. ASTM (Philadelphia), 1990 [revised annually]

6. INTERNATIONAL ORGANISATION FOR STANDARDISATION. *Soil quality sampling: Part 2: Guidance on sampling techniques*. ISO CD 10381-2. Committee Draft (April 1993)

7. LOVELL, J. Environmental samples and carefully controlled shipping to prevent degradation. *Pollution Prevention*, June, 1993, 59-61

8. BRITISH STANDARDS INSTITUTION. Draft for Development Code of Practice for the identification of contaminated land and its identification. BS DD 175: 1988. BSI (London), 1988

9. NEDERLANDS NORMALISATIE-INSTITUTE. *Soil: Investigation strategy for exploratory survey*. NVN 5740. NNI (Delft), 1991

10. INTERNATIONAL ORGANISATION FOR STANDARDISATION. *Soil quality – sampling: Part 1: Guidance the design of sampling programmes*. ISO CD 10381-1. Committee Draft (December 1993)

11. INTERNATIONAL ORGANISATION FOR STANDARDISATION. *Soil quality – sampling: Part 4: Guidance on the procedure for the investigation of natural and near – natural and cultivated soils.* ISO CD 10381–4. Committee Draft (November 1993)

12. INTERNATIONAL ORGANISATION FOR STANDARDISATION. *Soil quality – sampling: Part 5: Guidance on the procedure for the investigation of soil contamination of urban and industrial sites.* ISO CD 10381-5. Committee Draft (October 1993)

13. SMITH, M.A. and ELLIS, A.C. An investigation into methods used to assess gasworks sites for reclamation. *Reclamation & Revegetation Research*, 1986, **4**, 183-209

14. SIEGRIST, R.L. and JENSSEN, P.D. Evaluation of sampling method effect on volatile organic compound measurements in contaminated soils. *Environmental Science and Technology*, 1990, **24**, 1387-1392

15. HEALTH AND SAFETY EXECUTIVE. *Disposal of explosives waste and the decontamination of explosives plant.* HS (G) 36. HMSO (London), 1987

16. NAYLOR, J.A., ROWLAND, C.D. and BARBER, C. *The investigation of landfill sites.* Technical Report TR 91. WRC (Medmenham), 1978

17. CRIPPS, J.C., BELL, F.G. and CULSHAW, M.G. (eds). *Groundwater in engineering geology.* Geological Society (London), 1986

18. FREEZE, R.A. and CHERRY, J.A. *Groundwater.* Prentice Hall (Englewood Cliffs, New Jersey), 1979

19. NIELSEN, D.M. (ed.). *Practical Handbook of Groundwater Monitoring.* Lewis Publishers (Chelsea, MI), 1991

20. CLARK, L. *The field guide to water wells and boreholes.* Geological Society (London), 1988

21. BRITISH STANDARDS INSTITUTION. *Water Quality: Part 6: Sampling: Section 6.1 Guidance on the design of sampling programmes.* BS 6068: Part 6: Section 6.1: 1981 (confirmed 1990) [ISO 5667/1-1980]. BSI (London), 1991

22. BRITISH STANDARDS INSTITUTION. *Water Quality: Part 6 Sampling: Section 6.2 Guidance on sampling techniques.* BS 6068: Part 6: Section 6.2: 1991 [ISO 5667/2 – 1991]. BSI (London), 1991

23. BRITISH STANDARDS INSTITUTION. *Water Quality: Part 6 Sampling: Section 6.11 Guidance on sampling of groundwaters.* BS 6068: Part 6: Section 6.11: 1993 [ISO 5667/11 – 1993]. BSI (London), 1993

24. BRITISH STANDARDS INSTITUTION. *Water Quality: Part 6 Sampling: Section 6.3 Guidance on the preservation and handling of samples.* BS 6068: Part 6: Section 6.3: 1986 [ISO 5667/3 – 1985]. BSI (London), 1986

25. YOUNG, P. *Representative groundwater sampling and an overview of surface water sampling, sample handling and storage.* Course notes: Site Investigation, Centre for Extension Studies, Loughborough University, November 1992

26. UNITED STATES ENVIRONMENTAL PROTECTION AGENCY. *Handbook: Groundwater Volume II.* USEPA (Washington, DC), 1991, EPA/625/6-90/016b

27. BRITISH STANDARDS INSTITUTION. *Water Quality: Part 6 Sampling: Section 6.4 Guidance on sampling from lakes, natural and man made.* BS 6068: Part 6: Section 6.4: 1987 [ISO 5667/4 − 1987]. BSI (London), 1987

28. BRITISH STANDARDS INSTITUTION. *Water Quality: Part 6 Sampling: Section 6.6 Guidance on sampling of rivers and streams of samples.* BS 6068: Part 6: Section 6.6: 1991 [ISO 5667/6 − 1991]. BSI (London), 1991

29. BRITISH STANDARDS INSTITUTION. *Water Quality: Part 6 Sampling: Section 6.12 Guidance on sampling of sediments.* BS 6068: Part 6: Section 6.12: 1993 [ISO 5667/12 − 1993]. BSI (London), 1993

30. WATER RESEARCH CENTRE/NATIONAL RIVERS AUTHORITY. *Pollution potential of contaminated sites: a review.* R & D Note 181. WRC (Medmenham), 1993

31. TAYLOR, L. Laboratory analysis techniques surveyed. *Pollution Prevention*, 1993, **3**, (5), 56-58

32. MONTGOMERY, R.E., REMETA, D.P. and GRUENFELD, M. Rapid on-site methods of chemical analysis. In: *Contaminated Land: Reclamation and Treatment.* M.A. Smith (ed.). Plenum (London), 1985, pp 257-310

33. SIMMONS, S.M. (ed.) *Hazardous Wastes Measurement* Lewis Publishers (Chelsea, MI), 1990

34. AMERICAN CHEMICAL SOCIETY. *Proceedings of the Eighth Annual Waste Testing and Quality Assurance Symposium*, Arlington, Va. ACS, 1992

35. US ENVIRONMENTAL PROTECTION AGENCY *et al. Proceedings of the Second International Symposium 'Field screening methods for hazardous wastes and toxic chemicals'.* US EPA (Las Vegas) 1991

Further reading

BAYNE, C.K., SCHMOYER, D.D. and JENKINS, R.A. Practical reporting times for environmental samples. *Env. Sci. Tech.* 1994, **28** (8), 1430-1436

KEITH, L.H. Throwaway data. *Env. Sci. Tech.* 1994, **28** (8), 389A-390A

5 Risk assessment

5.1 INTRODUCTION

Risk assessment and risk reduction together comprise the overall process of **risk management.** There is an overlap between **risk assessment** (comprising hazard identification and assessment, risk estimation and risk evaluation) and **risk reduction** (comprising risk evaluation and risk control).

The options available for **reducing/controlling** risks are described in Volume IV. This Section focuses on risk assessment and describes :

- The basic objectives and scope of risk assessment
- The concepts and definitions used in risk assessment
- The types of data required and methods currently used to assess human health and other risks associated with the presence of contamination
- The use of risk assessment in contaminated land applications
- The communication of risk information.

The discussion that follows focuses on chemical contamination and risks to human health, but the principles involved are equally applicable to other forms of contamination, hazards and targets. Risks can also be viewed in a wider context. For example, the risk could relate to a failure to comply with regulations (possibly damaging a hitherto good record), adverse effects on site infrastructure, damage to existing or planned 'investment', negative impacts on community or public relations, or the costs of remedying damage caused by contamination. Such potential risks have an important bearing on whether, and what type of remedial action is eventually taken, and their place in the risk assessment process should not be overlooked.

The particular requirements of ecological assessments are discussed in Appendix 12.

5.2 OBJECTIVES AND SCOPE

5.2.1 Main objectives

In contaminated land applications risk assessment should be an ongoing process used both to inform the design, and interpret the findings, of site investigation (see Section 1). The main objectives of risk assessment are to :

- systematically determine whether there are currently any risks to human health or other targets (e.g. flora and fauna, the water environment, built environment), and whether, if such risks exist, they are acceptable
- determine the effects of foreseeable events, such as weather extremes, rising water table, flooding, increase in neighbouring populations etc., on the nature and magnitude of the risks

- determine the consequences (e.g. potential impacts on the environment, groundwater resources, public health) of a change of use, development, redevelopment or other works on the site

- identify the critical contaminants and associated factors (e.g. pathways) relevant to the site so that the steps necessary to reduce risks to 'acceptable' levels, both currently and in the foreseeable future, can be determined

- make judgements about the significance and acceptability of identified risks

- help to set objectives and priorities for reducing risks

- provide a rational and defensible basis for discussing a proposed course of action with third parties (e.g. regulators, communities, funders, insurers etc.).

5.2.2 Scope of risk assessment

Risk assessment comprises four main components.

1. **Hazard identification** – identifying the hazards that may be associated with a particular site or group of sites.

2. **Hazard assessment** – assessing the degree of hazard associated with a site or group of sites (what type and how much of the hazard could be available and reach a target) through consideration of plausible hazard/pathway/target scenarios.

3. **Risk estimation** – estimating the likelihood that an adverse effect will result from exposure to the hazard and the nature of the effect. Risk estimation may focus on human health effects, effects on flora and fauna, the water environment or other targets such as building materials.

4. **Risk evaluation** – evaluating the significance of estimated risks, taking into account available guidelines and standards, the uncertainties associated with the assessment and the costs and benefits of taking action to mitigate the risks.

The process for a simple site-specific risk assessment is illustrated in Figure 5.1.

For a risk to exist there must be a hazard (a property or situation capable of causing harm), a pathway and a target: if any of these are absent there is no risk. The hazard/pathway/target scenario must also be plausible. Thus for a contaminant to present a risk, it must be potentially harmful (e.g. toxic), there must be a pathway (e.g. migration in groundwater) along which the contaminant can travel, and there must be a target (e.g. a human consumer of the contaminated groundwater).

For example, for human health (toxicological) risks three conditions must be satisfied:

1. The source of the hazard must be toxic, mobile and in a bioavailable form.

2. A transport mechanism capable of conveying the hazard to a potentially sensitive target must be available.

3. A realistic target exposure route (e.g. ingestion), must exist.

Consideration of such 'plausibility criteria' at the hazard assessment stage can lead to early elimination of non-critical hazard/pathway/target scenarios from subsequent more detailed assessment.

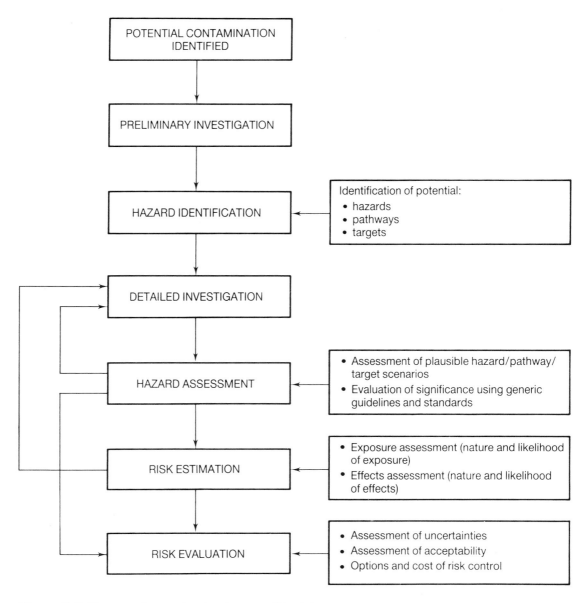

Figure 5.1 *Process for a simple site specific risk assessment*

A risk (i.e. the probability that the adverse effect associated with a hazard will occur) can be expressed in either quantitative or qualitative terms. The ability to quantify a risk depends on the availability of data on the contributory factors, incidence and outcome of similar past events. Such data are not readily available in a contaminated land context, therefore it is often difficult to provide quantified estimates of risk. Moreover, the cost of obtaining the data needed to provide a quantified estimate of risk may not be justified where a qualitative assessment provides an adequate basis on which to proceed.

In practical terms, quantified estimation of risks will rarely be required for typical development projects on contaminated sites. In these cases, hazard assessment (using generic guidelines, such as the ICRCL trigger concentrations[1,2]) or qualitative risk estimation using other relevant criteria (see Section 6 for a discussion of these) will often be sufficient to make decisions on appropriate courses of action.

A full site-specific risk assessment leading to a quantified estimate of risks is most likely to be required:

- when generic values for contaminant(s) are not available or not appropriate (e.g. in relation to the water environment)
- when there is benefit e.g. cost savings) to be gaiined by justifying acceptance of concentrations above generic values (which can be shown to be over-conservative)
- where the frequency of exposure and/or dose is likely to be significant
- in respect of contaminated sites currently used for 'sensitive' purposes, such as housing, where actual current risks must be assessed
- in respect of other identified 'problem' sites where there appears to be an actual or imminent threat to public health or the environment
- in respect of operating sites where the owners/operators or potential purchasers wish to assess current and future liabilities.

Whether or not risk estimates are quantified, the important point to note is that risk assessment provides a systematic framework for the routine scrutiny of the data being collected and assessed, and opportunities to make regular judgements about:

- the need for further information
- the uncertainties associated with the assessment
- the extent to which a fully quantified estimation of risks is either required or justified.

5.2.3 Main uses of risk assessment

Risk assessment can be used on a site-specific basis to:

- determine whether remedial action is required
- establish the standard of remediation required to control or reduce risks to an acceptable level
- evaluate different risk control/reduction options
- demonstrate to third parties (e.g. site owners, financial institutions, regulator) that a proposed form of action is the best way to proceed.

Risk assessment can also be used to classify and prioritise sites for further investigation or remedial action according to the risks they currently present or may pose in the future.

5.3 CONCEPTS AND DEFINITIONS

The term 'risk assessment' is a loosely applied term commonly used to cover a wide range of assessment procedures from simple and generalised statements on possible hazards and risks, to formalised procedures described here as comprising:

- hazard identification
- hazard assessment
- risk estimation
- risk evaluation.

The use of hazard and risk assessment procedures requires an understanding of a number of concepts and terms. For convenience these are listed in Box 5.1 under the main elements of risk assessment although an understanding of all the concepts and terms is essential before commencing risk assessment.

Only the principal components of risk assessment are considered in detail here. Descriptions of the remaining elements can be found in Appendix 8.

Box 5.1 *Concepts and terms used in risk assessment*

Hazard identification and assessment

- hazard
- hazard identification
- pathway
- target

Risk estimation

- human health risk
- environmental risk
- safety
- toxicity
- exposure assessment
- exposure and dose
- dose-response relationship ecotoxicity

Risk evaluation

Note that 'risk assessment' and related terms are used and defined in different fields of activity, contexts and countries, and by different interests ('stakeholders') in different ways.

The terminology and guidance provided here and throughout the Report has been developed by reference to a number of sources including:

- guidance produced by the US Environmental Protection Agency (US EPA)[3-7] (note, however, that the terminology employed here differs from that used by the US EPA),
- publications produced by the Health and Safety Executive[8,9], the Royal Society Study Group on risk assessment (RSSG)[10], the Institution of Chemical Engineers[11] (note that the IChemE suggests that 'risk assessment' should be used only when there has been a quantified estimation of the likelihood of events), and the Engineering Council[12].

Hazard

A hazard is a *property or situation* (e.g. the presence of a toxic contaminant, accumulation of potentially explosive gases) that has the *potential* to cause harm (e.g. personal injury, damage to property) to a defined target or target group (e.g. occupiers of a site, neighbouring property). In a wider context, a hazard may also be defined in financial, social or compliance (regulatory) terms.

Risk

Risk is the chance or probability that an adverse effect will occur under certain specified conditions. The risk may be to humans or other targets. The risk may be expressed in quantitative terms, with values ranging from (a nominal) zero (expressing certainty that 'harm' will not occur) to 1 (indicating certainty that it will). In many cases, however, it may be possible to describe the risk only in qualitative terms e.g.:

- high, medium, low, negligible
- chronic or acute
- present or future
- confined to certain sensitive targets
- dependent on certain events occurring (e.g. flooding).

Examples of potential human health risks associated with contaminated land are listed in Box 5.2.

Some sources (e.g. the Royal Society[10]) include the magnitude of the consequences of outcome in the definition of 'risk' but in the assessment of contaminated sites this 'consequence analysis' is usually, and is here, considered to be part of the risk evaluation stage to be executed and described separately from the statement of probability.

Box 5.2 *Examples of potential human health risks in contaminated land applications*

- Acute (short-term) risks to the workforce coming into close contact with hazardous substances. Consideration must be given to any attendant physical hazards (which themselves represent a major category of acute risks) since these may enhance the risk of harm from exposure to the substance; for example a cut may provide a ready pathway for substances to enter the body
- Chronic (long-term) risks to workers arising from contact with hazardous substances including, for example, potential carcinogens where damage may not materialise until some time after the exposure event
- Acute risks to members of the public, or other targets, in close proximity to the site due to large scale events, such as fire or explosion; the release or migration of potentially toxic gases and vapours; or direct contact with contaminants (e.g. during trespass)
- Chronic risks to nearby and remote populations, and other targets, due to the release of contaminants from the site over long time periods
- Acute risks to users of the site
- Chronic risks to users of the site.

Target

Targets or receptors for contamination include humans (either in general or a specific group such as small children or workers), fauna or flora, ecosystems, the water environment, a valuable resource (e.g. potable water source, commercial or recreational fishery, mineral deposit).

Hazard identification

This is the systematic identification of the hazards that may be associated with the site. Examples of hazards include the presence of potentially toxic or carcinogenic substances; explosive or flammable gases or other materials; other physical hazards such as used syringes or glass; pathogenic organisms (e.g. anthrax); and corrosive substances and solvents that can degrade or otherwise damage the building fabric or site services.

Initially hazards are identified by preliminary site investigation (desk study, site reconnaissance and initial exploratory work). The findings can then be used to design subsequent phases of investigation to ensure they are properly focused on collecting the right type and level of information to fully characterise hazards, pathways and targets, and assess risks (see Section 5.4). Hazard identification data can also be used to prioritise sites for further action (see Section 5.7).

Hazard assessment

During hazard assessment, site investigation data are first of all assessed (e.g. by comparison with background or reference levels of contamination — see Section 6) to establish whether a hazard exists. Note that in some circumstances a contaminant present at concentrations below background may still represent a hazard. Possible hazard/pathway/target scenarios should also be reviewed for plausibility; a contaminant may be present at above background concentrations and still not constitute a hazard if it cannot come into contact with a target.

Sufficient site investigation data (including information on the contaminants, the geological and hydrological setting of the site, and the distribution and nature of the potential targets associated with the site) must be available to properly characterise hazards, pathways and targets.

Examples of the types of hazards, pathways and targets which may be relevant are shown in Figure 5.2.

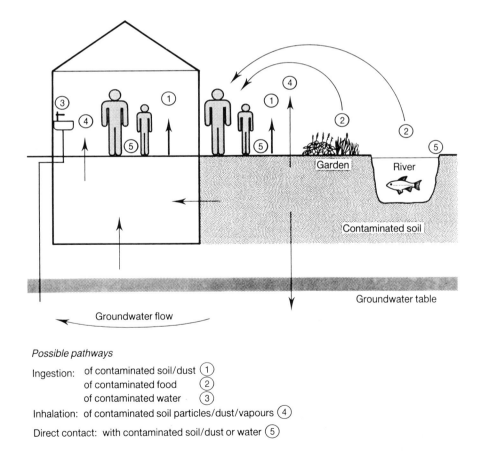

Figure 5.2 *Example of possible exposure pathways*

For plausible scenarios, the concentrations of contaminants found on the site can then be compared with (acceptable) concentrations that are judged not to be harmful to the target. Acceptable concentrations may be available in the form of generic guidelines (e.g. the ICRCL trigger concentrations for soils) or standards (e.g. Environmental Quality Standards for air or water). Care should be taken to ensure that the guidelines or standards used are properly representative of the hazard/pathway/target scenario being assessed (note for example that the ICRCL threshold trigger concentration values for copper, nickel and zinc relate to potential phytotoxic, rather than human toxicity effects − see Section 6).

Depending on the outcome, the hazard assessment may indicate that:

- observed levels of contamination are unlikely to pose a risk to specified targets and no further action is required

- further investigation and/or assessment (perhaps involving site-specific risk estimation) is needed before the significance of observed levels of contamination can be properly judged

- levels of contamination are such that there is no doubt as to the need for remedial action.

Hazard assessment does not provide a quantified estimate of risks although it can be used to prioritise or rank risks. The assessment stops at the point of potential consequence and by-passes the detailed risk estimation stage of risk assessment. The acceptability of the contamination is judged solely against the assumptions built into the generic guideline or standard.

Risk estimation

Risk estimation is the process of estimating the probability that an unwanted 'event' will occur under certain defined conditions. Both quantitative (e.g. numerical index) and qualitative (e.g. narrative) estimates may be provided. Ability to provide a quantitative estimate of probability requires that data are available on the incidence and outcome of similar past events, together with detailed information on site conditions.

Risk estimation involves :

- an *exposure assessment* − estimation of the exposure of the target to the hazard, taking into account the frequency and magnitude of the exposure, timescales and target characteristics

- an *effects assessment* − a consideration of the effect of the exposure on the target. Effects assessments may focus on human health (*toxicity assessment*), impacts on flora (e.g. phytotoxic effects) and fauna, water quality, ecosystems, building materials or other defined targets.

For risk estimation detailed information must be available on the :

- nature, concentration, and anticipated behaviour of the contaminant(s) in the environment

- the pathway(s) by which the contaminants(s) may come into contact with the target

- the nature of the exposure to the target (e.g. through inhalation, ingestion, direct contact)

- the effect that the contaminant may have on the target under defined conditions

- the range of factors which may affect any or all of these parameters thereby influencing the magnitude and nature of the risk.

Since direct monitoring and time-dependent data describing the behaviour of the contaminant in the environment or the response of the target to the hazard are unlikely to be adequate/available, models will usually be required to estimate risks. Particular care should be taken to ensure that models are representative of the situation being assessed and that the assumptions used are valid.

An important feature of risk estimation is that it enables *action values* (i.e the point at which further assessment or remedial action should be taken) and *remedial values* (e.g. the residual contaminant concentrations which any remedial action must achieve) to be determined on a site-specific basis.

Risk evaluation

Risk evaluation involves a consideration of :

- quantitative or qualitative statements about the risks associated with the contaminants found on the site

- other site-specific factors which may affect the nature of the risks (e.g. climatic changes, exposure of previously buried material, propensity to flooding, construction/redevelopment activity such as piling)

- uncertainties in the quantified risk estimates (if any have been made)

- the significance and acceptability of the estimated risks in relation to the current and likely future situation

- the costs and benefits of taking action to control or reduce unacceptable levels of risks.

A decision to take action to deal with a contaminant depends on the judgements made during risk evaluation about the significance of the risks. Both the magnitude and the consequences of the risks should be taken into account. For example even where the risk (e.g. that a fire or explosion will occur on the site) is judged to be low, the consequences of such an event (in terms of human health effects, damage to property etc.) may be considered so unacceptable that action is taken to reduce or eliminate the risks. Conversely a risk which is considered to be high (e.g. that a small proportion of young trees on a landscape scheme will die due to phytotoxic effects) may be tolerated in the light of wider health and environmental considerations, and the cost of taking measures to prevent the loss (i.e. to reduce the risk).

Because of the inherent uncertainties usually associated with risk assessment, 'worst case', as opposed to 'typical' or 'average', scenarios are sometimes employed to build in adequate margins of safety. Box 5.3 gives an example of the factors considered in a conservative risk assessment[13] concerning the presence of dioxin. Uncertainties are typically addressed using assumptions, for example on the nature and impact of the exposure and effects. Box 5.4 lists examples of typical uncertainties and the assumptions used to estimate the health risks associated with contaminated land.

It is essential to identify and document all the assumptions used since an important task of risk evaluation is to critically review uncertainties to establish how realistic the assumptions are in practice, and how sensitive the risk estimation is to any changes in the assumptions used.

The output from a risk evaluation process should therefore include statements on :

- the magnitude of the risks and nature of effects

- the uncertainties involved and their sensitivity to the assumptions made

- how changes in the assumptions might change the acceptabilities of identified risks

- where risk reduction efforts should be directed to achieve most effect, given the existing and possible future circumstances of the site.

> **Box 5.3** *Assumptions used to assess risks associated with presence of dioxin in soils*
>
> Assumptions which might be made in calculating the theoretical risk to a child living on a site where garden soils, plants and neighbouring land are contaminated with dioxin (e.g. due to aerial deposition):
>
> The child
> - consumes a teaspoon of contaminated soil per day
> - plays regularly in the garden
> - consumes fish from a nearby river
> - is supplied with drinking water from the river
> - swims in the river
> - consumes fish in higher than 'average' amounts
> - eats vegetables primarily from the family garden
> - drinks milk from cows grazing on contaminated fodder

> **Box 5.4** *Examples of uncertainties and assumptions used to estimate human health risks*
>
> Typical uncertainties:
> - extent, concentration and chemical form of contaminants
> - behaviour of contaminants in the environment e.g. effects of chemical reaction, degradation, attenuation, dilution, adsorption, dispersion etc.
> - exposure routes and duration (e.g. acute vs chronic)
> - short and long-term effects under specified exposure conditions
> - variability of human response in terms of age, gender and general health characteristics
> - effect of exposure to more than one substance simultaneously
>
> Typical assumptions:
> - typical exposures are to highest observed concentrations of contaminants
> - all or most of the material is biologically available
> - low levels of attenuation, degradation etc. occur along the exposure pathway
> - exposure assessment based on maximally exposed and most vulnerable individual(s)

5.4 INFORMATION REQUIREMENTS

Two main types of information are needed to conduct risk assessment:

1. Site characterisation data including information on the behaviour of the contaminant in the environment.

2. Information on the toxicological and other properties associated with contaminants.

Information requirements are highly site-specific, depending on the nature of the contamination and the nature of the targets. Tables 5.1 to 5.3 provide check lists of factors which should be addressed when identifying and defining hazards on contaminated sites. Table 5.4 provides a classification of some of the organic compounds which may be encountered. A further classification of chemicals in relation to their properties (e.g. volatility) is provided in Appendix 1 of Volume VII.

Table 5.1 *Substances and materials which may be a hazard if present on a site*

General category	Examples
Toxic, narcotic and otherwise harmful gases and vapours	Carbon dioxide, carbon monoxide, hydrogen sulphide, hydrogen cyanide, toluene, benzene.
Flammable and explosive gases	Acetylene, butane, hydrogen sulphide, hydrogen, solvents, petroleum hydrocarbons
Flammable liquids and solids	Fuel oils, solvents, process feedstocks, intermediates and products
Combustible materials	Coal residues; ash; timber; domestic, commercial and industrial waste
Materials liable to self-ignition	Paper, grain, sawdust if present in large volume and sufficiently damp to initiate microbial degradation
Corrosive substances	Acids and alkalis, reactive feedstocks, intermediates and products
Zootoxic metals (and their salts)	Cadmium, lead, mercury, arsenic, beryllium, copper
Other zootoxic chemicals	Pesticides, herbicides
Carcinogenic substances	Asbestos, arsenic, benzene, benzo(*a*)pyrene, vinyl chloride
Mutagenic substances	Vinyl chloride, trichloroethylene,
Teratogenic substances	Arsenic, ethylene glycol, 1,3 butadiene, diphenyl amine
Allergenic substances and sensitisers	Nickel, chromium
Substances causing skin damage	Acids, alkalis, phenols, solvents
Phytotoxic metal	Copper, zinc, nickel, boron
Reactive inorganic salts	Sulphate, cyanide, ammonium, sulphide
Pathogenic agents	Anthrax, polio, tetanus, Weil's disease
Radioactive substances	Some hospital laboratory wastes, radium contaminated objects and wastes, some mine ore wastes, some non-ferrous slags or phosphorus slags
Physically hazardous materials	Glass, hypodermic syringes
Vermin	Rats, mice, cockroaches

Table 5.2 *Possible hazards, pathways and targets*

Aspect	Example
Possible hazards	Flammable substances, explosives, asphyxiants, toxic substances, allergens, pathogens, carcinogens, mutagens, teratogens, sensitizers
Possible pathways for humans	Direct contact, inhalation (e.g. dust, gases, vapours), ingestion (e.g. soil, surface water, drinking water), indirect ingestion (e.g. eating contaminated plants or animal products), absorption through skin or cuts etc., indirect inhalation (e.g. volatile organics 'stripped' from shower water)
Possible targets	Site workers, future occupiers or users, neighbouring occupiers and users, soil quality, surface and groundwater quality, ambient air quality, flora and fauna, buildings and services, mineral resources

More comprehensive information on the contaminants typically associated with particular types of sites is available in the DOE/BRE industry profiles.[14] Further details on hazards can be found in the CIRIA guidance document on the health and safety aspects of contaminated sites[15] (see also the reference sources listed in Appendix 1). Information on the distribution, fate and effects of selected chemicals in the environment are given in a series of documents prepared by the Department of the Environment[16].

Table 5.3 *Characteristics of substances*

Characteristics	Examples	Characteristics	Examples
Chemical:	Composition Speciation Reactivity Flammability Persistence Lipid solubility Partition coefficients	Biological:	Toxicity Corrosivity Pathogenicity Carcinogenicity Mutagenicity Teratogenicity Genotoxicity Allergenicity
Physical:	Volatility Solubility in water Solubility in non-aqueous phases Fineness Viscosity	Presentation in the environment:	Phase: solid/liquid/gas Quantity Concentration Accessibility Mobility

Table 5.4 *Classification of some organic chemicals*

Class of compound	Examples
Pesticides	Inorganic and organometallic types, substituted compounds containing phosphorous, nitrogen and halogen groups
Aliphatic hydrocarbons	Low molecular weight hydrocarbons, mineral oils
Aromatic hydrocarbons	Benzene, toluene, xylene
Polycyclic aromatic hydrocarbons (PAHs)	Naphthalene, pyrene, fluoranthene, anthracene
Substituted aromatic compounds	Pentachlorophenol, polychlorinated biphenyls (PCBs), polychlorinated dibenzo dioxins and furans
Substituted aliphatic compounds	Trichloroethane, tetrachloroethylene, brominated compounds

5.4.1 Site characterisation data

Box 5.5 illustrates the type of site characterisation data needed to conduct a site-specific risk assessment. The example relates to a 'typical' contaminated site which may pose a risk to a nearby potable water source. The example illustrates the importance of:

- collecting the right types of data and information (i.e. directed at the targets potentially at risk)
- collecting sufficient data to fully describe hazards, pathways and targets
- ensuring data are of the right quality
- ensuring that information describing the dynamics of the risks are available (e.g. because land adjacent to a site is likely to be developed, or a drinking water supply is established where none existed before)
- providing benchmarks (for example on actual or anticipated variations) so that the degree of uncertainty of the risk estimation can be properly understood.

> **Box 5.5** *Information needed to assess risks presented by a contaminated site to potable groundwater source*
>
> **The problem** A contaminated site may pose a risk to a potable groundwater source
>
> **Information requirements**
> Distribution and concentration of the contaminants (soils, surface waters and groundwater)
>
> Chemical form of the contaminant
>
> Physical properties of contaminants (e.g. density, solubility)
>
> Physical properties of soils and other site materials (e.g. particle size distribution)
>
> Measure of the 'availability' of the contaminant to the environment (e.g. leachability)
>
> Factors affecting availability (pH, clay content, organic matter content) and likely variation
>
> Rainfall
>
> Ground surface permeability
>
> Ground permeability at site (e.g. for different strata or materials) and along travel pathway(s)
>
> Geochemistry of aquifer (e.g. cation exchange capacity)
>
> Groundwater levels and variations
>
> Direction and velocity of groundwater flow
>
> Location of water courses, ponds, etc
>
> Location and nature of groundwater/surface water interfaces (e.g. springs, streams)
>
> Location of 'target' abstraction point
>
> Information on nature of abstraction point (depth, screened length)
>
> Abstraction rates and dilution factors
>
> Location, nature, and utilisation factors of any other abstraction wells
>
> Concentration of contaminant in abstracted water
>
> Information on health, and other, effects of the contaminant
>
> Size and location of the exposed population
>
> Size of the population likely to be exposed in the future
>
> **Assumptions** Assumptions will include those relating to direct exposure to the contaminant in drinking water, and other possible routes such as food preparation, bathing and washing etc.
>
> The total exposure of the target to the contaminant must be taken into account to fully estimate risks.

5.4.2 Toxicological and other data

A range of parameters (e.g. LD_{50}, Acceptable Daily Intake, etc.) are available that provide some guidance on the 'safety' or otherwise of particular substances. These can be used to derive 'safe' levels of exposure using appropriate uncertainty (safety) factors. In other cases, guidance on the acceptable concentration of a substance in the media of concern may be obtained by reference to standards or guidelines for air or water quality since these are themselves usually defined in terms of the concentration of a substance presenting no practicable risk of harm to exposed populations. Standards on water and food are also well developed. Those for air, and particularly for soil are less so. Media specific standards and guidelines can be used to derive site-specific action and remediation values, or used as inputs to the generation of generic values (see Section 6).

Important reference sources for information on the toxic, and other hazardous properties of substances in relation to human health are given in Table 5.5.

Table 5.5 *Information sources on the impact of hazards to humans*

Nature of the hazard	Mode of action	Source (criterion)
Toxicity/carcinogenicity	Ingestion, inhalation, absorption	HSE (e.g. OESs, MELs) DOE (e.g. IRPTC, WMP) WHO (e.g. ADIs) US NTP US NIOSH US DHHS/ATSDR US EPA (e.g. IRIS, HEASTs) US NCI
Corrosivity	All routes	US NIOSH HSE
Combustibility	--	DOE (FRS)
Flammability/ explosiveness	--	HSE DOE (FRS, WMP) CIRIA
Asphyxiation	Inhalation	HSE

Key:

CIRIA	= Construction Industry Research and Information Association
DOE (IRPTC, WMP)	= Department of Environment, (International Register of Potentially Toxic Chemicals, Waste Management Papers)
DOE (FRS)	= Department of Environment (Fire Research Station)
HSE (OESs, MELs)	= Health and Safety Executive (Occupational Exposure Standards, Maximum Exposure Limits)
US NIOSH	= US National Institute for Occupational Safety and Health
US NTP	= US National Toxicology Program
US DHHS/ATSDR	= US Department of Health and Human Services/Agency for Toxic Substances and Disease Registry
US NCI	= US National Cancer Institute
USEPA (IRIS, HEASTs)	= US EPA (Integrated Risk Information System, Health Assessment Summary Tables)
WHO (ADIs)	= World Health Organisation (Acceptable Daily Intakes)

5.4.3 Evaluation of data quality

Before commencing risk assessment it is essential to check that sufficient base data are available, for example on :

- the nature, extent and variability of the contamination

- the nature, extent and variability of the contamination reaching the targets, or target locations, and at intermediate points

- the exposure pathways (e.g. in relation to geology, hydrology, meteorological conditions)

- other site-specific factors relevant to the interpretation of the risk, including for example, topography, climate, socio-economic factors, current use(s) and planned use(s)

- actual or anticipated toxicological or other effects.

It is pointless attempting to carry out a formal quantitative risk assessment if the database is insufficiently extensive or of inadequate quality.

Deficiencies can occur in all five areas outlined above. For example, failure to detect the presence of a highly mobile, persistent and toxic compound negates the assessment process, as would the failure to recognise that a potentially harmful substance has already reached its target. Failure to identify a critical pathway (e.g. sand lenses in predominantly clay strata) could lead to an underestimation of the potential for the migration of contaminated water. Failure to identify plans for the development of a large housing estate next to the site, would lead to an underestimation of the size of the population potentially at risk. Failure to recognise the carcinogenic potential of a contaminant can lead to the use of inappropriate guidelines or standards for comparative purposes, an invalid effects assessment and erroneous conclusions on the significance of the risks.

The quality of data used in risk assessment is important in respect of both the source and the target location, particularly the latter. In quantitative risk estimation, a number of uncertainty factors or assumptions will be used to trace the contaminant from the source to the target which, collectively, may outweigh the variations in analytical data obtained. This does not, however, obviate the need to know how reliable the data are since even small errors in determined values, or failure to detect a contaminant at the target location, can have profound effects on the estimation of risk.

In the absence of definitive data relating to certain aspects of risk assessment, assumptions are used to produce a numerical estimate of risk. In these cases, it is essential to state explicitly the basis of the assumptions, and whether assumptions or other data limitations have the effect of over, or under, estimating the degree of risk associated with the site. Sensitivity analysis should be performed.

5.5 CONDUCTING A SITE-SPECIFIC RISK ASSESSMENT

The following Section provides a brief introduction to the procedures and techniques used to assess the human health risks associated with contaminated land. Reference is made to US protocols (but note that they may not necessarily be directly applicable in the UK). Procedures are illustrated by an example shown in Box 5.9. Supporting information on human health assessment can be found in Appendix 9. Note that risk assessment is a specialist activity which should only be undertaken by appropriately qualified personnel.

Hazard identification and assessment

Using data collected during site investigation, a qualitative evaluation is made to determine whether any of the contaminants identified at the site have the potential to cause adverse effects on health and, if so, which are likely to be the most important. Flow charts of the type shown in Figure 5.3 are helpful in identifying critical hazards/pathways/targets, and deciding whether a generic or site-specific approach should be used. Table 5.6 shows how critical and non-critical scenarios can be summarised in matrix form.

Note that several other hazard/pathway/target scenarios may be relevant in individual cases, e.g. inhalation of contaminated dusts, ingestion of contaminated fish (from polluted lakes or rivers), contravention of water quality standards etc.

Where appropriate, critical scenarios are assessed further in the risk estimation stage.

Table 5.6 *Matrix of hazard/pathway/target scenarios*

Contaminated medium	Exposure route	End use of site			
		A	B	C	D
Soil	Ingestion	✓	✓	×	×
	Dermal contact	✓	✓	×	×
	Inhalation (of volatiles)	✓	✓	✓	✓
Groundwater	Ingestion	×	×	×	×
	Dermal contact	×	×	×	×
Surface water	Ingestion	×	×	×	×
	Dermal contact	×	×	×	×

Note: End uses

A Residential
B Recreational
C Recreational with protective cover
D Light industrial (e.g. use class B1 in England and Wales)
✓ Likely to be significant
× Unlikely to be significant

Source: from reference 17

Risk estimation

Exposure assessments are carried out to estimate the type and magnitude of exposure by the target to the hazard. Exposure assessments can be made using simple one-dimensional models that relate concentrations of contaminants in the source (e.g. soil), intakes of the contaminated medium by the exposed individual, exposure frequency and duration, and averaging time.

A range of different models can be used to inform the exposure assessment (see Section 5.6). In all cases it is essential that all assumptions used, including the characteristics of the exposed individual, are explicitly stated. An example of an exposure uptake equation is shown in Box 5.6[18].

Once the exposure assessment has been completed, a toxicity assessment is carried out to identify the effects of the exposure and to provide, where possible, an estimate of the likelihood that adverse health effects will occur as a result of the exposure.

In most cases this will involve comparing a calculated intake by the target with relevant standards or guidelines. For example, a calculated intake of a hazardous substance can be compared to a WHO Acceptable Daily Intake value. Analogous values can be derived for other types of potential exposures, e.g. a value based on EC drinking water standards (for intakes via a contaminated water supply source); Occupational Exposure Standards (for an airborne hazard); or a potentially explosive mixture of a gas in air (for an explosion hazard).
Box 5.7 summarises the protocols used in the USA. These may not be consistent with UK approaches to cancer estimation.

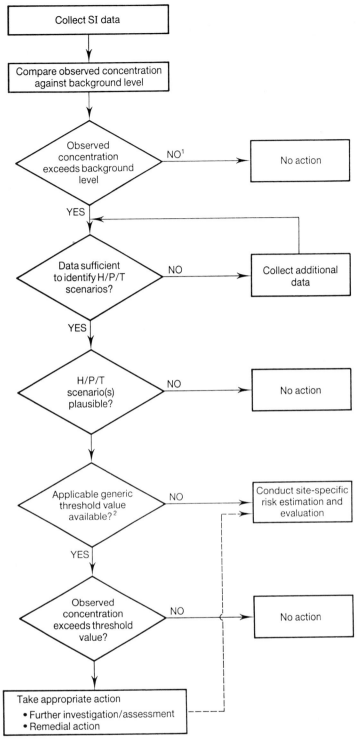

KEY

1. Extreme care should be exercised when comparing observed concentrations against background values (see Section 6)
   SI = Site Investigation
   HPT = Hazard/Pathway/Target
2. Applicable to plausible scenarios only

Figure 5.3 *Hazard assessment using generic guidelines and other criteria (single contaminant only)*

> **Box 5.6** *Example of an exposure uptake equation for ingestion of contaminated soil*
>
> **Chemical: Zinc**
> **Effects : Non-carcinogenic**
>
> **Equation: Intake** (mg/kg-day) = (CS × IR × CF × FI × EF × ED) / (BW × AT)
>
> $$= 1095 / 273750 = 4 \times 10^{-3}$$
>
> **Input variables:**
>
> CS = Concentration of chemical in soil (mg/kg) = 1000
> IR = Ingestion rate (mg soil/day) = 100
> CF = Conversion factor = 10^{-6}
> FI = Fraction ingested from source (unitless) = 1
> EF = Exposure frequency (days/year) = 365
> ED = Exposure duration (years) = 30
> BW = Body weight (kg) = 25
> AT = Averaging time (period over which exposure is averaged, days)
>
> **Assumptions:**
>
> CS : Maximum concentration (worst case scenario)
> IR : Based on age groups > six years old (US data)
> CF : 10^{-6} (kg/mg)
> FI : Conservative assumption
> EF : Maximum exposure (worst case scenario)
> ED : National (US) upper bound time at one residence
> BW: Based on age groups between 6 and 9 years old (US data)
> AT : Obtained by multiplying Exposure Duration (30 years) × Exposure Frequency (365 days)
>
> This example makes a number of conservative assumptions in calculating the potential uptake by a child through ingestion. For example, it assumes that ingestion of 100 mg soil (equivalent to 0.1 mg of zinc) occurs daily, that this occurs over a 30 year period and that the body weight remains at 25 kg for the entire period. In some cases such assumptions are required by the authorities laying down the relevant protocols or requiring the assessment to be made; in some cases they will appear as 'default values' in computerised models. Unless there are protocols requiring the use of particular assumptions, the assessor must decide what assumptions should be made in carrying out a site specific estimation of risk. The reasons for the judgements made must be explicitly stated. Part of the 'uncertainty analysis' carried out at the risk evaluation stage is concerned with testing the sensitivity of the risk estimate to changes in the assumptions used in such calculations.

> **Box 5.7** *Toxicity assessment in the United States*
>
> In the US formalised toxicity assessment procedures have been developed which use chronic reference doses (RfDs) and cancer slope factors (CSFs) to evaluate the relative toxicity of hazardous substances. The RfD can be regarded as the maximum daily intake of a compound that can occur without adverse health effect, even in sensitive individuals such as children. The CSF is an *upper bound* estimate of the probability of excess cancer per unit of intake averaged over the lifetime of an individual.

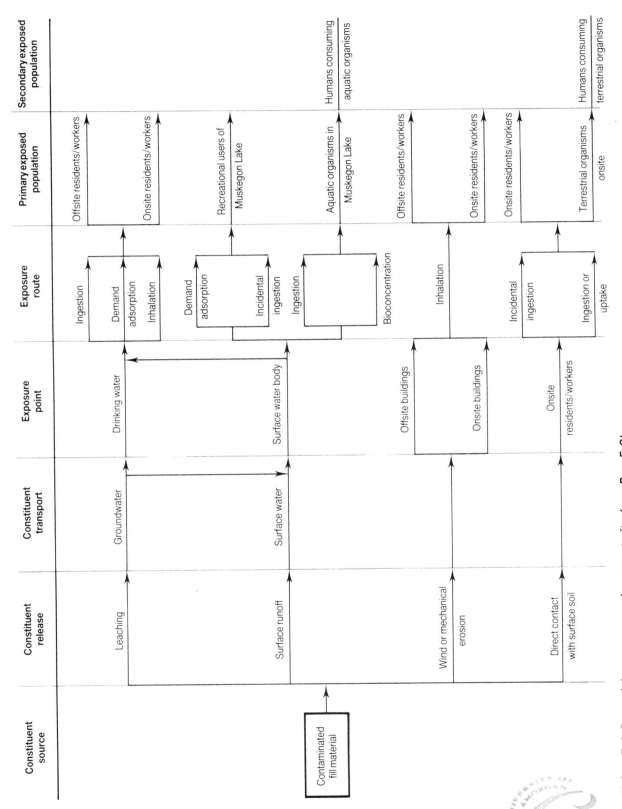

Figure 5.4 *Potential exposure pathways at site (see Box 5.9)*

CIRIA Special Publication 103

Risk evaluation

In risk evaluation all the information obtained during the risk assessment is assembled and developed into qualitative or quantitative expressions of risk. Judgements are then made about the acceptability or otherwise of the risk taking the assumptions used and uncertainties of the assessment into account.

The results of the comparison between calculated values and relevant standards and guidelines are an important part of the evaluation. For example, if a calculated intake is lower than the ADI it can be assumed that for all practical purposes exposure to the hazardous substance, under the conditions specified, is unlikely to result in any significant adverse health effects in the individual concerned. If the calculated value exceeds the ADI, judgements have to be made about the likelihood of adverse effects and whether, if the effects did occur, they would be acceptable.

Whatever the result it is essential to review the assessment to ensure that the assumptions used are reasonable and that the conclusions reached are fully justifiable. Sensitivity to changes in the assumptions should be tested.

In the US guidance is given on the acceptability or otherwise of calculated intakes as shown in Box 5.8. Note that these criteria may not necessarily be acceptable in the UK for all types of

---

Box 5.8 *Acceptability of hazards and risks in the United States*

In the US two different measures of acceptability are used depending on the nature of the hazards:

Non-carcinogenic hazards

For non-carcinogenic hazards the calculated intake is divided by the relevant RfD (see Box 5.7) to obtain a Hazard Index (HI). If the HI < 1, exposure under the conditions specified is considered unlikely to result in significant health effects for the individual concerned.

Hazard indices greater than 1 indicate that adverse effects may occur, and judgements have to made as to the likelihood and significance of the risks.

Carcinogenic hazards

For carcinogenic hazards the calculated intake is multiplied by the corresponding CSF (see Box 5.7) to obtain an estimate of the increased risk of developing cancer over a lifetime through exposure to the hazardous substance under the conditions specified.

For USEPA Superfund sites, the acceptable range of increased cancer risk is between 1×10^{-4} and 1×10^{-6} (but note UK HSE criteria for residents living close to major hazard plant, see text).

Multiple exposures

Where exposure to a number of different substances occurs hazard indices and carcinogenic risk values for individual substances are simply added to obtain a cumulative value. These values are regarded in the same way as substance-specific values as far as acceptability is concerned. Thus a cumulative hazard index of less than 1, or a cumulative cancer risk within the specified range, would still be regarded as acceptable.

Note that some substances can have both non-toxic and carcinogenic effects, and may therefore make a contribution to both the Hazard Index and Carcinogenic Risk Value.

risks. For example, the HSE criteria for risks (of receiving a dangerous dose not just risk of death) for residents living in the vicinity of major hazard plant[8] are 10^{-5} to 5×10^{-7}. In practice judgements about the acceptability or otherwise of a quantified risk estimate depends on a wide variety of additional factors. For certain sites an estimated risk lower than 10^{-6} may not be acceptable for a variety of social and economic reasons, while a hazard index of 1.5 may be considered acceptable at a different site for the same reasons.

Box 5.9 shows the application of a quantified risk estimation to a potential redevelopment site. In this case it was possible to show that development for residential purposes would result in lower risks to human health than leaving the site in a derelict condition.

5.6 USE OF MODELS IN RISK ASSESSMENT

In quantitative risk assessment two types of models are usually employed:

- models describing the behaviour of contaminants in the environment
- models describing known toxicological, or other effects, on human health or other targets.

Some models combine elements of both approaches for standardised scenarios.

Models belonging to the first category aim to quantify the movement of contaminants along the travel pathway or exposure route. They may be hydrological models (see also Volume VI) that describe contaminant migration in response to surface or groundwater movement; they may be air emission and dispersion models describing the effect of wind action on a source of solid or gaseous contaminants (see below).

Models belonging to the second group are derived from toxicological, or other, studies of the substances of concern. They generally focus on the dose-response relationship. Different types of models are used to describe non-carcinogenic as opposed to carcinogenic effects.

The advantage of using models in risk assessment is that they provide a formal and consistent framework for risk estimation based on well documented effects for the contaminant under consideration. Most also have the facility to alter input parameters to more closely reflect site-specific conditions. Modelling techniques (e.g. Monte Carlo simulations) are now being developed that allow variability in the value of input data to be taken into account when estimating risks[19]. However, the underlying assumptions used can vary and different models can give quite different results[19]. They may also be difficult to validate in practice. Where possible, site assessment data should be used as inputs when estimating exposures and dose, rather than relying on default or predicted values. Examples where the direct use of monitoring data may not be possible include :

- where exposure points are remote from monitoring points (e.g. air dispersion modelling)
- where time dependent data are unavailable (e.g. advection of contaminants in a groundwater plume)
- where monitoring data are unavailable because of detection limits (e.g. where a contaminant is released into a large body of water and is subject to large dilution factors).

In this last case, although adverse toxicological effects could in theory occur at contaminant concentrations below the detection limit, the output of any modelling at this level needs careful and rigorous scrutiny, especially if it is the main factor behind any decision to take remedial action.

> **Box 5.9** *Case study example of a quantified risk estimation (after Traves[18])*
>
> In 1990 a subsurface investigation was carried out on a vacant industrial site in Muskegon, Michigan prior to its redevelopment for 36 condominiums. A risk assessment was carried out to evaluate potential risks associated with current use (mainly due to trespass) and future residential use.
>
> Three exposure pathways were identified as being critical to the two land use scenarios:
>
> - ingestion and dermal absorption of contaminated soil
> - dermal absorption and incidental ingestion of contaminated surface water from Muskegon Lake
> - ingestion of contaminated fish caught in the lake.
>
> The full range of possible exposure pathways is illustrated in Figure 5.4
>
> Exposure point concentrations were estimated to calculate the potential uptake of the compounds of concern. For the current land use scenario, exposure point concentrations were based on the concentrations detected during the site investigation. For the future residential development scenario, exposure point concentrations were based on conservative fate and transport modelling because clean backfill (0.3–2.7m) was to be placed on the site prior to development.
>
> Results from the USEPA Toxicity Characteristic Leaching Procedure (TCLP) were used to calculate the concentrations of copper, cyanide, lead, zinc and benzene likely to be transported back into the clean backfill after placement. Surface water concentrations were estimated by modelling groundwater concentrations at the point of discharge into the lake, and then applying a conservative dilution factor of 10.
>
> Chronic daily intake (CDI) calculations (based on reasonable maximum exposure (RME)) for both a hypothetical adult and child were performed to provide greater sensitivity in the estimation of risks. Reasonable worse case assumptions were used to estimate CDIs. Based on the results of the CDI calculations, the compound with greatest adult and child CDI under the current property use scenario was TPH (total petroleum hydrocarbons) based on soil ingestion and dermal absorption. Under the future use scenario, zinc was the compound with greatest CDI based on soil ingestion and dermal absorption.
>
> A toxicity assessment was then performed using EPA tabulations of data (e.g RfDs) as the main reference source. Lack of data for some of the compounds present on the site left an element of uncertainty in the risk evaluation.
>
> Total hazard index and carcinogenic risk values were calculated for the two scenarios (see below). The total hazard index for a child under the current use scenario was estimated at 3000 times that for residential development and over 3 times the US EPA's acceptable figure. Total excess cancer risks were within the acceptable range for both uses, but were also less for the development scenario.
>
> | Exposed population | Site risk criteria (total) | Vacant land | Residential use |
> |---|---|---|---|
> | Adult | Hazard Index | 0.1 | 0.00007 |
> | | Excess cancer | 2×10^{-10} | 3×10^{-13} |
> | Child | Hazard Index | 3.0 | 0.0008 |
> | | Excess cancer | 2×10^{-8} | 3×10^{-11} |

Ferguson has described DOE sponsored research to develop a model that can generate both site-specific and generic values for contaminated sites[20]. A joint National Rivers Authority/Department of the Environment research study is also developing a modelling tool to assess the risks to the water environment from new landfill developments (completion is expected in 1995). Modelling to determine acceptable concentrations of radionuclides in soils has been described by Till and Moore[21]. Pollard *et al*[22] report the use of a fugacity model to describe the environmental fate of organic wood treatment chemicals following release into the environment (the fugacity of a gas is a measure of its ability to disperse). There is also a specialist literature on the environmental and health effects of soils contaminated by petroleum which includes the application of modelling techniques (see for example reference 23).

Appendix 10 gives examples of the use of environmental models in contaminated land applications.

5.7 USES OF RISK ASSESSMENT

Risk assessment can be used either :

- in relation to individual sites to identify site-specific action and remediation values
- as a means of prioritising individual sites (within a larger portfolio) for subsequent more detailed investigation and possible remediation.

5.7.1 Application to individual sites

Risk assessment can be used to :

- demonstrate that generic guidelines (where available) provide sufficient protection to be adopted as site-specific values (this assumes that the assumptions made in deriving such guidelines are known — see Section 6)
- derive site-specific (or more likely media and/or location specific) action and remediation values
- assist in the evaluation of different risk reduction/ remediation strategies
- provide a rigorous and rational basis for discussion with others (e.g. regulators, the local community) about a proposed course of action
- assist a consideration of long-term legal and financial liabilities to current and future land owners
- provide the basis for assessment and quantification of insurance risks.

Derivation of site-specific values

The derivation of site-specific action or remediation values (e.g. for the protection of human health) can be achieved by a process of back calculation following the same hazard/pathway/target relationships established by the risk assessment (for summary account of the process see references 24 and 25, and Box 5.10):

1. Calculate the total acceptable (human) exposure level from known dose-response relationships incorporating safety/uncertainty factors as necessary.

2. Determine the acceptable contaminant concentration at the point of exposure for each pathway.

3. Back calculate the maximum concentration permissable at the source.

4. Apply appropriate uncertainty/safety factors to derive action or remediation values.

> **Box 5.10** *Example of derivation of a site-specific remediation value*
>
> The remediation value is calculated using a contaminant fate and transport model in the backward mode. Such models take into account the attenuation of contaminants as they migrate from the source to the target. Attenuation processes include advection, dispersion, chemical and biological degradation, volatilisation and adsorption.
>
> The general equation is:
>
> $$RV = \frac{C}{AF}$$
>
> where:
>
> RV = remediation standard (e.g. in mg/kg, mg/L)
> C = receptor point concentration (mg/kg, mg/L in a simple case)
> AF = attenuation factor defined as the ratio of the receptor point concentration to the source concentration.
>
> The model assumes that receptor point concentrations are below solubility limits and saturated vapour pressure limits. However, this will be a characteristic of the model used, site dimensions and time frame, and can be determined for a range of source concentrations by running the model.
>
> C can be set as a defined standard (e.g. drinking water standard) or be calculated as an 'acceptable value' from consideration of dose-response relationships and Rfds etc (see Appendix 9).
>
> Where more than one contaminant is present risk or hazard indexes are combined using established criteria.
>
> The acceptable concentration at the point of exposure can be calculated by application of the general formula:
>
> $$C = \frac{(BW)\ (AT)\ (ADI)}{(CR)\ (EF)\ (ED)}$$
>
> where:
>
> C = contaminant concentration in ground water, soil or other media at the point of exposure (mg/kg, mg/L)
> ADI = acceptable chronic daily intake (e.g. mg/kg day)
> BW = body weight (kg)
> AT = averaging time (i.e the period over which exposure is averaged) (days)
> CR = contact rate (eg for drinking water L/day)
> EF = exposure frequency (days/year)
> ED = exposure duration (years)
>
> Special models to calculate site-specific standards have also been developed (see Section 6)

Selection of appropriate risk reduction strategies

The purpose of risk reduction is to remove the source of a hazard, to impede or eliminate actual or potential exposure routes/pathways, to remove existing targets or avoid creation of new targets. Several potentially useful options will usually be available on a site-specific basis.

A risk assessment approach automatically helps the selection of appropriate remedial methods and strategies by :

- providing a means of formally defining site-specific action and remediation values

- requiring the collection of detailed information on the nature of the contaminants and their behaviour in the environment — this gives a good insight into how contaminants are likely to behave during remediation or within a remedial system (e.g. containment system)

- defining critical pathways and targets — this helps to identify those which may be relevant during remediation as well as those created as a result of remedial action (e.g. through the generation of a more mobile or toxic by-product); it also aids understanding of the constraints that pathways and targets may place on the choice of remedial action.

Basis for acceptance

A well researched and presented risk assessment provides an excellent basis for discussion with other parties having a legitimate interest in the site. These include regulatory bodies, institutional investors, insurers, and the local community. It indicates that significant effort has been made to fully characterise the actual and potential risks presented by the site, and the best way of managing critical risks. Since the basis of the assessment, and in particular any judgemental aspects, must be fully documented, other parties are able to have full access to the decision-making framework. While this clearly provides opportunities for challenging the assumptions and judgements made, it also encourages better understanding of the risks involved and increases the chances that a consensus view will be reached.

5.7.2 Comparison of sites

A risk based approach can be used to classify sites in terms of the risks they currently present; for example hazard identification findings can be used to classify sites into those that :

- require immediate action to halt or reduce pollution which is causing, or threatens to cause harm in the near future, to human health, the environment or some other target

- require additional urgent investigation to more fully characterise the nature of the risks

- should be investigated further in due course and monitored in the interim

- do not appear to present any risk.

Such a classification might be made:

- following a site identification programme by a government or regulatory body (although not currently a feature of UK environmental law)

- by a property owner with extensive property holdings who wants to identify potential liabilities and set priorities for precautionary action

- by a potential purchaser of an industrial or commercial concern wishing to know where to concentrate pre-purchase investigations or those parts of a property portfolio to avoid.

Classification can be carried out at several levels, and at various stages on an iterative basis, according to the quality and quantity of available data. The base data can be obtained from :

- desk studies only

- preliminary investigations (desk studies and site reconnaissance work)

- hazard assessments

- qualitative risk estimation

- quantified risk estimation.

A classification which is not based on quantified risk estimates can be made more systematic by incorporating a *hazard ranking system*. In this approach, risk factors reflecting the nature of the hazards, the existence of pathways and the proximity of targets are combined to score a site in terms of hazards to human health and/or the environment (separate scorings are usually made). Hazard ranking systems can be simple or complex depending on the number of factors used, and the complexity (e.g. weighting of individual factors) of the scoring systems.

The terminology applied to this process around the world is not consistent. No clear distinction is made between systems that are officially described as 'hazard ranking' in one country and what might be termed a 'qualitative risk assessment', 'simple quantitative risk assessment' or 'preliminary risk assessment' in others. As noted below, the UK Department of the Environment is developing[26] what it terms a 'qualitative risk assessment procedure' that takes account of pathways and target vulnerability.

Important principles to be observed when developing and applying hazard ranking systems (and other preliminary assessment systems) include the need to :

- tailor the system to the task in hand (systems should be no more complex than is absolutely necessary)

- bear in mind at all times the arbitrary nature and uncertainties typically associated with scoring systems

- revise rankings as new information becomes available

- take social, economic and other non-technical factors into account when deciding priorities based on such a system.

Factors which might be taken into account by an organisation (e.g. local authority) carrying out an initial screening exercise are listed in Table 5.7.

Table 5.7 *Possible screening factors for potentially contaminated sites*

Factor	Example
Development status of the site	<ul><li>Currently being redeveloped (work in progress)</li><li>Already redeveloped for 'sensitive' use</li><li>Already redeveloped for less sensitive uses</li><li>Permission granted for redevelopment</li><li>Scheduled for redevelopment (e.g. in local plan)</li><li>Unlikely to undergo change of use</li><li>Undeveloped (this does not mean the site is uncontaminated since certain types of agricultural land, for example, may be highly contaminated)</li><li>Vacant/derelict</li></ul>
Locational factors etc	<ul><li>Current or historical use (suggesting that the site may be contaminated)</li><li>Current use (if residential whether houses, flats, sheltered accommodation etc.)</li><li>Number of people potentially at risk</li><li>Presence of particularly vulnerable groups (e.g. children, elderly)</li><li>Use of neighbouring land</li><li>Liability to flooding</li><li>Stability of water table (currently rising in a number of cities)</li><li>Signs of obvious damage to plants, water courses etc.</li><li>Age of buildings and services (e.g. is leakage adding to problems, is repair needed ?)</li><li>Ownership</li><li>Isolated site or one of a group</li><li>Whether the site is also subject to diffuse pollution</li><li>Ecological targets (including sensitive ecosystems) at risk</li><li>Proximity of surface water bodies</li><li>Proximity of water abstraction points (public, private, industrial)</li><li>Superficial and underlying geology</li></ul>

Traditionally hazard ranking systems have been used to prioritise sites for action according to their current condition, rather than to evaluate sites for development. However, they could be used for this purpose since they can provide an indication of the extent of any future 'threat' posed by the site either as a result of the development process itself or should an uncontrolled change of use occur (i.e. one without appropriate remedial works being carried out). For example:

- A local planning authority or other regulatory body might incorporate such an assessment into its forward planning strategy in order to identify industrial sites requiring particular attention if an application for a change of use is made.

- A developer might employ such a system to guide its land purchasing policies so that it can identify sites which might be particularly costly and time consuming because of the additional works likely to be required or the time needed to obtain appropriate approvals etc.

A developer might find such a system particularly valuable where responsibilities for land purchase or development planning are decentralised to ensure that projects are evaluated on a similar basis in different parts of the organisation. A more or less systematic assessment of non-technical factors could form the third strand of the overall ranking (the other two being human and ecological risk factors, see Appendix 12).

Documented hazard ranking systems include that used by the US Environmental Protection Agency to decide which of the sites brought to its attention under the 'Superfund' Programme should be placed on the National Priorities List for Federal Action.[27] However, a number of other countries (e.g. Norway[28], Denmark[29], Australia & New Zealand[30,31]) have developed more or less complex systems (see Volume XII, Policy and Legislation). Canada has recently introduced a system[32] under its national programme for the identification and treatment of contaminated land[33]. In the UK, British Gas is evaluating the use of a 'semi-quantitative risk assessment' process based on the Canadian system[32] to prioritise sites for further action[34].

The Department of the Environment has been preparing guidance on a qualitiative risk assessment procedure to categorise sites into relative risk categories. This procedure takes target vulnerability and pathways into account[26].

5.8 COMMUNICATION OF RISKS

The communication of information about health or environmental risks is an inherent part of risk management[35]. It may be required at any stage of the process and in different contexts. A number of different parties (stakeholders) with varying interests may be involved, including:

- regulatory bodies
- professional advisers
- financial organisations
- land and property owners
- the local community.

At any one time a party may have more than one interest in a site. The type and nature of interest will directly influence:

- the timescale of the response
- the degree of involvement of the different parties

- the degree of concern and perception of risk
- information requirements
- the methods and routes used to exchange information.

There is clearly potential for misunderstanding and ineffective communication of risk based information.

Failure in risk communication can lead to inappropriate and costly decisions. The full participation of all those who have to manage, as well as those who have to bear, the risks (to the environment, health, investment and finance) is essential. Experience also indicates that technical assessment procedures are likely to be ineffective unless the communication requirements of risk managers and bearers are taken fully into account.

REFERENCES

1. DEPARTMENT OF THE ENVIRONMENT. *Guidance on the Assessment and Redevelopment of Contaminated Land.* ICRCL Guidance Note 59/83 (2nd Edition). DOE (London), 1987

2. DEPARTMENT OF THE ENVIRONMENT. *Notes on the Restoration and Aftercare of Metalliferous Mining Sites for Pasture and Grazing.* ICRCL Guidance Note 70/90. DOE (London), 1990

3. UNITED STATES ENVIRONMENTAL PROTECTION AGENCY. *Risk assessment guidance for Superfund, Volume I: Human health evaluation manual (Part A).* USEPA Office of Emergency and Remedial Response (Washington DC), EPA/540/1-89/002, 1989

4. UNITED STATES ENVIRONMENTAL PROTECTION AGENCY. *Risk assessment guidance for Superfund, Volume II: Environmental evaluation manual.* USEPA Office of Emergency and Remedial Response (Washington DC), EPA/540/1-89/001, 1989

5. UNITED STATES ENVIRONMENTAL PROTECTION AGENCY. *Risk assessment, management, and communication of drinking water contamination.* USEPA Office of Water, (Washington DC), EPA/625/4-89/024, 1989

6. UNITED STATES ENVIRONMENTAL PROTECTION AGENCY. *The risk assessment guidelines of 1984.* USEPA Office of Health and Environmental Assessment (Washington DC), EPA/600/8-87/045, 1987

7. Guidelines for exposure assessment. *Notice Federal Register*, **57**, No 104 (May 1992), pp 22888-22938

8. HEALTH AND SAFETY EXECUTIVE. *Risk criteria for land-use planning in the vicinity of major industrial hazards.* HMSO (London), 1989

9. HEALTH AND SAFETY EXECUTIVE. *The tolerability of risk from nuclear power stations.* HMSO (London), 1988

10. ROYAL SOCIETY STUDY GROUP. *Risk : Analysis, perception and management.* Royal Society (London), 1992

11. INSTITUTION OF CHEMICAL ENGINEERS. *Nomenclature for hazard and risk assessment in the process industries*. ICE (Rugby), 1985

12. ENGINEERING COUNCIL. *Guidelines on risk issues*. (London), 1993

13. KEENAN, R.E. et al. Taking a risk assessment approach to RCRA corrective action. In *Proceedings of a Conference on Developing Clean-up Standards for Contaminated Soil, Sediment and Groundwater – How Clean is Clean ?* Water and Environment Federation (Alexandria, Virginia), 1992, 255-275

14. DEPARTMENT OF THE ENVIRONMENT. ICRCL Industry Profiles. DOE (London), in preparation

15. STEEDS, J.E., SHEPHERD, E. and BARRY, D.L. *A guide to safe working practices for contaminated sites*. Report 132. CIRIA. (London), in the press

16. DEPARTMENT OF THE ENVIRONMENT (TOXIC SUBSTANCES DIVISION). *Environmental hazard assessment: benzene*. Building Research Establishment (Garston) 1991

(Various other titles available, e.g. toluene, di-(2-ethylhexyl) phthalate and 1,1,1 trichloroethane).

17. EDULJEE, G. Application of risk assessment to contaminated land. In: *Proceedings of a Conference on Contaminated Land Policy, Economics and Technology*. IBC (London), 1993, Paper No.2

18. TRAVES, L. Applying risk assessment concepts to evaluate alternative uses of contaminated industrial properties. In: *Proceedings of a Conference on Decommissioning, Decontamination and Demolition: Closure of Factory Sites*. IBC (London), 1992, Paper No. 4

19. *Proceedings of a Conference on Developing Clean-up Standards for Contaminated Soil, Sediment and Groundwater: How Clean is Clean ?* Water and Environment Federation (Alexandria, Virginia), 1993

20. FERGUSON, C. and DENNER, J. Soil remediation guidelines in the United Kingdom: A new risk based approach. In: *Proceedings of a Conference on Developing Cleanup Standards for Contaminated Soil, Sediment and Groundwater: How Clean is Clean ?* Water and Environment Federation (Alexandria, Virginia), 1993, pp 205-212

21. TILL, J. E. and MOORE, R. E. A pathway analysis approach for determining acceptable levels of contamination of radionuclides in soil. *Health Physics*, **55** (3), 1988, 541-548

22. POLLARD, S. J. T., et al. Screening of risk management options for abandoned wood-preserving plant sites in Alberta, Canada. *Canadian Journal of Civil Engineering*, in press

23. CALABRESE, E. J. and KOSTECKI, P. T. *Soils contaminated by petroleum: environmental and health effects*. Wiley (Chichester), 1988 (first of a series of publications)

24. ANON. Risk assessment methods for deriving clean-up levels. *The Hazardous Waste Consultant*, May/June 1991, 1.1-1.6

25. STATE OF CALIFORNIA DEPARTMENT OF HEALTH SERVICES. *Technical standard for determination of soil remediation levels.* SCDHS, 1990

26. DEPARTMENT OF THE ENVIRONMENT. *Risk assessment and characterisation procedures for sites which may be contaminated.* DOE (London), in preparation

27. WEELS, S. and CALDWELL, S. Overview of the revised Hazard Ranking System (HRS). In: *Proceedings Superfund '90 Conference, Washington 1990.* Hazardous Materials Research Institute, Silver Spring, MD, 1990, pp 71-76.

28. FOLKESTAD, B. Approach to clean up standards in Norway. In: *Proceedings of a Conference on Developing Clean-up Standards for Contaminated Soil, Sediment and Groundwater: How Clean is Clean ?* Water and Environment Federation (Alexandria, Virginia), 1993, pp 187-194.

29. POULSEN, M. M., *et al.* Approaches to the development of soil and groundwater clean up standards in Denmark. In: *Proceedings of a Conference on Developing Clean-up Standards for Contaminated Soil, Sediment and Groundwater: How Clean is Clean ?* Water and Environment Federation (Alexandria, Virginia), 1993, pp 93-114

30. AUSTRALIAN AND NEW ZEALAND ENVIRONMENT AND CONSERVATION COUNCIL. *Australian and New Zealand Guidelines for the Assessment and Management of Contaminated Sites.* ANZECC & National Health and Medical Research Council, 1992

31. McFARLAND, R. Simple quantitative risk assessment technique for prioritisation of chemically contaminated sites in New South Wales, Australia. In: *Proceedings of a Conference on Risk Assessment.* Health and Safety Executive (London), 1992

32. CANADIAN COUNCIL OF MINISTERS OF THE ENVIRONMENT. *National classification system for contaminated sites* (final draft). CCME (Winnipeg), 1991

33. SCHMIDT, J. Contaminated land: a Canadian viewpoint. In: *Proceedings of a Conference on Contaminated Land Policy, Regulation and Technology.* IBC (London), 1991, Paper No.2

34. WALKER, P.L., MUNRO, S., HAWKINGS, C.L, and SHEPHARD, F.E. A proactive approach to managing an inherited problem — the application of risk assessment to contaminated land. In: *Preprints Symposium on Contaminated Land: From Liability to Asset,* Birmingham, 1994, pp 1-16.. Institution of Water and Environmental Management (London) 1994

35. PETTS, J.Dealing with contaminated land within a risk management framework. In: *Proceedings Conference on Contaminated Land: Policy, Risk Management and Technology, London 1994.* IBC Technical Services (London) 1994

Further reading

ANON. *Protocol for the health risk assessment and management of contaminated sites.* South Australian Health Commission (Adelaide), 1991

ANON. *The health risk assessment and management of contaminated sites.* South Australian Health Commission (Adelaide), 1991

DENNER, J. Contaminated land: a framework for risk assessment. In: *Proceedings Conference Contaminated Land: Developing a Risk Management Strategy.* IBC (London) 1994

MYERS, K., VOGT, T. and WALES, J. Hazard ranking criteria for contaminated sites. *Land Contamination and Reclamation,* 1994, **2** (1), 13-18

PETTS, J. *Environmental impact assessment for waste treatment and disposal facilities.* Wiley (Chichester) 1994

6 Guidelines and standards

6.1 SCOPE

During risk assessment, use can be made of published lists of concentrations of contaminants in soil, water or other media that indicate that there should be no unacceptable risks to specified targets through specified exposure pathways, or that indicate the need for certain actions to be taken (e.g. further investigation, a quantitative risk assessment, remedial action).

There is no agreed national or international terminology for these values and terms such as 'soil quality criteria' and 'standards' may have different meanings in different countries and jurisdictions. Uniform and consistent usage of terms by people with different professional backgrounds and experiences is difficult to achieve. Great care is therefore needed to ensure that the policy, legal and technical basis of any numerical values used for assessment/remedial purposes is understood, and that the terminology is properly defined. The UK Department of the Environment has not published guidance on terminology.

For the purposes of this Report it has consequently been necessary to develop a terminology to describe a number of important concepts.

Two terms have been employed:

- *guidelines*
- *(quality) standards*

to describe lists of numerical values promulgated by governments or their agencies.

Guidelines and standards are defined as numerical *values* (e.g. mg/kg contaminant in soil) used for two main purposes:

1. To assess the significance of observed levels of contamination through the application of risk assessment.

2. To provide a measurable indicator of what is to be achieved by remedial action (i.e. a quantified remediation objective, expressed, for example, in terms of an 'acceptable' residual concentration of a contaminant).

The term '*value*' is used as a collective term for guidelines and standards and for their expression in numerical terms. This is typically a concentration of a contaminant in a specified medium (e.g. mg/kg in air-dried soil, mg/l groundwater, mg/m^3 in air).

The distinction made between guidelines and standards is as follows:

1. *Guidelines* are values to be applied with the aid of professional judgement.

2. *Standards* are limits made binding through government legislation or regulation which must be observed (within the appropriate regulatory framework) in all cases where they are applicable.

This distinction corresponds to the terms 'guideline value' and 'legally binding value' defined in the ISO standard *Soil quality — Vocabulary — Part 1: Terms and definitions relating to soil protection and soil pollution*[1]. Note that this distinction is not consistently made either in the available literature or in practice.

Numerical limits may also be made legally binding (usually on a site-specific basis) through private contract or application of relevant legislation (e.g. contracts of sale, contractually binding specifications, planning agreements, and planning conditions).

The inherent ambiguity of the term 'standard' is recognised[2,3] but its use here is analogous to its use in the composite term 'environmental quality standard' described in Box 6.4. It is assumed that the reader is well accustomed to its variable usage and will understand it in the context in which it is used.

Guidelines are usually produced by government or other regulatory agencies for general use, but may also be produced by other organisations for use on a corporate level.

The guidelines and standards issued by authoritative bodies are typically generic in nature. They usually incorporate risk factors which mean they can be used in a general capacity, irrespective of site-specific variation. While there is no legal obligation to adhere to guideline values, there is little scope for failing to comply with a generic standard.

In contaminated land applications three different sets of values are important:

1. *Dedicated guidelines and standards* (e.g. values derived specifically for the purpose of assessing the risks associated with contaminated land — they may be set for soil, groundwater, surface water or other media).

2. *Non-dedicated guidelines and standards* (values not derived specifically for the purpose but which nevertheless may be useful during risk assessment).

3. *Site-specific values* (values derived to reflect the specific circumstances of an individual site — see Section 5 for information on how these may be derived).

Both dedicated and non-dedicated values can be used depending on the situation. In all cases, however, it is essential to understand the policy, legal and technical basis of the values and any limitations on their use in particular applications.

This Section addresses:

- the types of standards and guidelines available for use
- the application of guidelines and standards to contaminated land
- dealing with variability in the application of guidelines and standards
- the guidelines and standards currently available in the UK.

Information on the guidelines and standards available overseas is provided in Appendix 11.

6.2 TYPES OF GUIDELINES AND STANDARDS

6.2.1 Dedicated guidelines and standards

Dedicated guidelines have been produced specifically to assist in the *assessment* of contaminated land in the United Kingdom, and as guidelines or standards in a number of other countries. Values (e.g. concentrations of contaminants) applicable to soils and, in some cases, groundwaters are available. They can be applied directly provided the assumptions used in their derivation are understood and the values are applicable to the situation being assessed. Although most generic values have been produced for the purposes of assessing observed levels of contamination they can be 'adopted' as site-specific remediation objectives (see Section 6.2.3 and Volume IV).

Most countries and jurisdictions supporting active site identification and remediation programmes have either promulgated their own values or adopted (with appropriate modification) those used in other countries[4].

The 'trigger values' provided by the UK Interdepartmental Committee on the Redevelopment of Contaminated Land (ICRCL)[5] are generic guidelines. They are purely advisory and have no status in law. The Dutch values are more akin to generic standards since they were introduced to support specific legislation on contaminated land (see Appendix 11 and Volume XII).

6.2.2 Non-dedicated guidelines and standards

These are values (a concentration in a specified medium) which, although not produced specifically for the purposes of assessing contaminated land, may nevertheless have an important bearing on both the assessment and remediation of contaminated sites. A wide range of non-dedicated guidelines and standards may be relevant (see Table 6.1 and Section 6.6).

Non-dedicated values are used in three main ways.

1. During risk assessment (see Section 5) to assess whether:

    - a hazard exists

    - adverse effects are likely to occur as a result of exposure to contaminants under defined conditions.

2. In the derivation of site-specific action and remediation values (see Section 6.2.3).

3. In the selection, design and implementation of remedial strategies where compliance with regulatory requirements (e.g. limits on the concentrations of contaminants in an effluent discharge or off-gas) is essential.

It is important to be aware which non-dedicated standards might apply during remediation since they may impose constraints on the use of particular remedial methods in certain situations (see Volume IV). For example, (i) unavoidable atmospheric emissions that might arise during excavation could exceed limits set for occupational exposure, and (ii) concentrations in air extracted during soil venting might require use of a gas treatment system rather than dispersion to atmosphere.

Table 6.1 *Examples of non-dedicated guidelines and standards*

Area of relevance	Examples
Health hazards (see also Section 5)	EC Drinking Water Standards WHO Acceptable Daily Intakes DOE guidance on toxic hazards for specific purposes (e.g. WMP 23) HSE Occupational Exposure Standards and Maximum Exposure Limits NRPB guidance on radiation hazards USEPA guidance on toxic hazards (e.g. HEASTs)
Fire/explosion hazards	HSE guidance on explosive and flammable materials DOE (WMP No. 27) guidelines on landfill gas
Hazards to the natural environment	Background concentrations in soils Environmental Quality Standards and Objectives (air and water)
Hazards to the built environment	DOE (Building Research Establishment) guidance on the protection of concrete in aggressive soils

6.2.3 Site-specific values

Site-specific values include :

1. Action values.

2. Remediation values.

Action values are *assessment criteria* that may be applied to all principal media (air, soil, water, biota and abiotic materials). The desirable/required action may take the form of:

- (additional) investigation or assessment

- emergency procedures to halt on-going pollution

- the preparation and execution of a longer-term remediation plan

- repair or maintenance work on a remedial system to rectify an actual or potential failure/deterioration in performance.

Remediation values define the performance to be achieved during, or by remedial works. In the first instance remediation values are defined (in either quantitative or qualitative terms) as *contamination related objectives* (see Volume IV). Examples of quantified contaminated related objectives include :

- that the level of residual polychlorinated biphenyls present in soil treated by incineration shall not exceed 0.5 mg/kg

- that the cumulative excess lifetime risk of cancer for children presented by a treated site shall not exceed one in one million.

Examples of qualitative objectives are :

- that the site shall be suitable for the construction of houses with 'large' gardens

- that the lake shall be capable of use for recreational fishing, swimming and sailing.

Qualitative objectives for the natural environment may be expressed in terms of preserving or enhancing amenity and aesthetic values. Qualitative objectives should be supported where possible by objective measurement that can provide evidence that a certain standard has been reached or maintained, e.g. simple colorimetric measurements of water quality, standardised measures of ecological quality and diversity. Qualitative objectives of this type should not be confused with the 'environmental quality objectives' or 'environmental quality standards' set by authoritative bodies.

During the process of selecting and evaluating remedial methods (see Volume IV) the options available for achieving stated objectives are assessed and a preferred remedy selected.
Typically a set of contamination related objectives are identified and a means of achieving them agreed with the regulatory authorities. This agreement may be essentially informal or may be made legally binding through private contract, 'planning agreements', or planning conditions.
Site-specific values can be obtained in one of three ways :

- by adopting a generic value without modification (e.g. an ICRCL threshold concentration value for free cyanide in soil assuming residential use of the site)

- by modifying a generic value

- by deriving values on the basis of a risk assessment.

The derivation of risk-based values may be necessary when :

- dealing with a complex site

- where directly applicable generic values are not available (e.g. when considering the impact of contaminated soil on the water environment)

- where the cost of establishing a less restrictive site-specific value is less than the cost of meeting a more restrictive generic value.

6.3 THE USE OF GUIDELINES AND STANDARDS FOR ASSESSMENT

6.3.1 Important terms and concepts

Important terms and concepts in the application of guidelines and standards for assessment purposes are listed in Box 6.1.

Box 6.1 *Important terms and concepts*

- contamination
- pollution
- background or ambient concentration
- reference concentration
- environmental quality objective
- environmental quality standard
- multi-functionality
- guideline vs standard
- generic vs site-specific

6.3.2 What is contamination?

An understanding of the terms *contamination* and *pollution* as generally applied in the UK is important in the application of guidelines and standards for assessment purposes. The distinction between contamination and pollution was made by the Royal Commission on Environmental Pollution[6] and has since been used from time to time by government. The definitions of pollution in recent UK legislation (e.g. Environmental Protection Act 1990) and in European Community Directives should be noted (see Box 6.2).

For the purposes of this Report the following definitions have been adopted:

Contamination is: the presence in the environment of an alien substance or agent, or energy, with a potential to cause harm.

Pollution is: the introduction by man into the environment of substances, agents or energy in sufficient quantity or concentration as to cause hazards to human health, harm to living resources and ecological systems, damage to structure or amenity, or interference with legitimate uses of the environment

Thus contamination (the presence of an alien substance) does not necessarily lead to pollution (an adverse environmental effect arising from contamination).

Note that the terms contamination and pollution are applied to both the process of contaminating/polluting and the results of that process.

Box 6.2 *Contamination and pollution*

s21(3) & (4) Environmental Protection Act 1990

'Pollution of the environment' means pollution of the environment due to the release (into any environmental medium) from any process of substances which are capable of causing harm to man or any other living organisms supported by the environment.

'Harm' means harm to the health of living organisms or other interference with ecological systems of which they form a part and, in the case of man, includes offence caused to any of his senses or harm to his property; and 'harmless' has a corresponding meaning.'

Directive 76/464/EEC

Pollution is defined as:

'the discharge by man, directly or indirectly, of substances or energy into the aquatic environment, the results of which are such as to cause hazards to human health, harm to living resources and aquatic ecosystems, damage to amenities or interference with other legitimate uses of water.'

Note that the terms 'contamination' and 'pollution' are not used on an equivalent basis in all countries and contexts: many countries do not make the distinction and some view contamination as being worse than pollution. The International Organisation for Standardisation (ISO) Standard *Soil Quality – Vocabulary – Part 1: Terms and definitions relating to soil protection and soil pollution* does not contain a definition of either term. However, it does urge all users of such terms to take care to define their meaning in the context in which they are being used.

6.3.3 Does contamination exist ?

An important first step in risk assessment is deciding whether the site is actually contaminated. This involves comparing observed levels of contamination with naturally occurring concentrations of substances. For soils, data on both *background* and *reference* concentrations are available (see Box 6.3).

> **Box 6.3** *Background and reference values for soils*
>
> *Background* or *ambient* concentrations provide a basis for judgements on whether contamination exists, or whether observed concentrations are atypical of the geographical setting. Figures are typically expressed in terms of *mean, modal, range,* and *typical* values. They may relate to national, regional or local situations. Background data are available on naturally occurring contaminants (e.g. cadmium, lead) in both undisturbed environments and those affected by human activities such as farming or urbanisation (see references 7 and 8). Data on chemicals of solely anthropogenic origin (e.g. pesticides) are more limited.
>
> Care should be exercised when comparing observed concentrations of contaminants with *local* background concentrations: these may well be above natural concentrations (e.g. as a result of aerial or water borne deposits or application of sewage sludge to land), or they may be 'high' because of local geological settings. In some cases background concentrations may be at a level at which harm to humans or other biota may occur. Background levels are therefore not necessarily 'safe'.
>
> *Reference* concentrations are values derived from a consideration of background concentrations (where data exist), soil properties and the characteristics of contaminants. They are particularly relevant in relation to contaminants that do not occur naturally. They are regarded as those concentrations of a contaminant which pose no unacceptable risks to potential targets. Reference values have been developed in the Netherlands, in support of a policy of restoring soils to multi-functionality (see Appendix 11).

> **Box 6.4** *Environmental quality standards and objectives*
>
> **Values for air**
>
> In the UK environmental quality values for ambient air have been set for suspended particulates, ozone, sulphur dioxide, nitrogen dioxide and lead. Both limit and guideline values are available. Values have been set at levels intended to be protective of both human health and the environment.
>
> **Values for surface water**
>
> Environmental quality standards (EQSs) are concentrations not to be exceeded at particular locations in the medium in question. Typically EQSs are set by a regulatory authority in such a way that the appropriate Environmental Quality Objective (EQO) for the medium as a whole can be achieved. In the UK values are set for different uses of water, e.g. values are defined for waters abstracted for drinking or agricultural purposes, or for the protection of certain forms of aquatic life (e.g. sensitive freshwater or marine species)[12]. Statutory criteria for classifying the quality of rivers were established in 1994 and it is expected that criteria for other types of controlled waters will follow in due course.

For air and water, environmental quality objectives and standards can be used to indicate whether observed concentrations of contaminants are likely to be unacceptable in the medium in question (see Box 6.4) but note that these will have been set above natural background concentrations and thus will not give a direct indication of whether contamination is present.

It is essential to understand the regulatory context when attempting to apply environmental quality objectives and standards to contaminated land situations.

In practice, background concentrations vary from place to place depending on the underlying geology. Tabulations of typical 'national' concentrations are a useful starting point.[7-9] Where data are available on both 'uncontaminated' and 'suspect' soils within a locality, statistical techniques are available to determine whether the area of concern is, in fact, contaminated.[7,10,11]

Differentiation between background concentrations of, and contamination by, organic compounds tends to present greater difficulties for three main reasons:

- although generally derived from human activity a number of the compounds of concern (e.g. polyaromatic hydrocarbons (PAHs), some halogenated hydrocarbons) do arise 'naturally' and have 'always' been present in the environment to some degree

- some organic compounds have become widely distributed in the environment

- there is a general shortage of data on background concentrations of organic compounds whether these arise naturally or through human activity.

In practice, there may be no need to establish that contamination exists provided guideline values are available which define that concentration of a substance presenting no additional risk either for a specified land use or irrespective of land use.

6.3.4 Does the contamination matter ?

Risk assessment is concerned with deciding whether observed concentrations of contaminants actually matter. Dedicated generic values have been developed to enable assessors to judge the significance of observed levels of contamination without necessarily having to resort to site-specific assessments in each case.

For soils the following situations can be identified (see also Figure 5.3 in Section 5) — (similar contamination categories can be also be developed for water or other media):

1. Observed concentrations of a contaminant are below, or are consistent with, typical ranges to be found in uncontaminated soils: the soil is *'uncontaminated'*.

2. Observed concentrations are above 'background' or 'reference values' but are at a level which do not increase the risks to a specified target by an unacceptable amount: the soil is *'contaminated but presents no unacceptable additional risks'* to the specified target (note: other targets may be at risk). This value is sometimes termed the *threshold value*.

3. Observed concentrations are above a level at which some increase in risk is anticipated and an assessment is needed to determine the extent of these risks and whether any action is required (such action may take the form of additional site investigation and some form of estimation and evaluation of risks): the soil is *'contaminated and may present additional risks'* to specified targets.

4. Observed concentrations are such that some form of action is essential either to overcome immediate problems or to avoid future problems arising from a change of use: the site is *'contaminated and presents an unacceptable risk'* to defined targets (e.g. to children playing on the site or to underlying groundwater or adjacent surface water because the soil contains soluble contaminants). The site is almost certainly *'polluted'* according to the definition given above.

This simple categorisation is of course complicated when, as discussed in 6.3.3, naturally occurring background levels are already above a concentration (limit value, critical load) at which harm is likely, or is actually causing harm, to specified targets. The soil is not contaminated or polluted in the terms defined in Section 6.3.2 as no alien substances are present, nevertheless they may present an unacceptable risk and depending on circumstances may require an appropriate response.

Most of the dedicated generic guidance developed by government organisations have used this type of analysis to identify numerical values signifying different levels of risk and prompting different types of responses (actions) from the assessor.

Various assumptions and uncertainty factors are used to derive the numerical values. In the case of threshold values care is taken to ensure that sufficient protection will be afforded the target if it is decided not to take remedial action in response to observed levels of contamination. Reference 13 shows how generic values for contaminants typically found on former gas works sites have been identified for the UK.

Since the risk characteristics of a contaminated medium vary according to its use, different assumptions and uncertainty factors can be used to develop a range of generic values for different uses. The concept of variable end use has been applied to the development of dedicated generic guidelines in a number of cases: UK trigger concentrations are defined for different uses of the site; Canadian criteria are available to cater for the different uses of soils and groundwaters associated with the site. In the Netherlands only one set of values is available because the policy context of multi-functionality states that soils and groundwater should be assessed according to their ability to perform any function (see Box 6.5).

6.3.5 Generic *vs* site-specific quantified assessment of risks

The main advantages of a generic approach to assessment include:

- consistency in the interpretation of site investigation findings
- ease of application
- relatively modest demands for site characterisation, toxicological and other data, and specialist expertise.

The main disadvantages of a generic approach include the possibility that [14,15]:

- a relevant generic value may not be available for the application under consideration
- generic values may be over or under protective for a particular application
- generic guidelines might be used in an inappropriate context because the assessor has not taken into account the risk factors used to derive the values

The derivation of risk based generic values assumes that the wide range of variables which determine the behaviour of contaminants in the environment and the responses of targets to exposure, can all be predicted in a reasonably standardised way. It also assumes that contaminants of concern can be reliably measured in the environment (either in absolute or relative terms as appropriate) and that adverse effects can be measured with a high degree of accuracy and precision. At the very least, this requires that regulations prescribing generic guidelines or standards should also specify the use of standard methods of investigation, sampling and analysis.

Clearly it is not possible to develop generic values that cater for the risk implications of all possible combinations of contaminants, media types, exposure pathways and targets. This means that there is sometimes an unavoidable need to conduct site-specific quantified risk assessments.

> **Box 6.5** *Multi-functionality of soils and waters*
>
> All soil and water bodies are capable of performing a variety of functions i.e. they are multi-functional, but few are omni-functional because an ability to perform one function may compromise the ability to perform others. Important soil and water functions include :
>
> **Soils**
>
> - Controlling element and energy cycles as a compartment of the ecosystem
> - Supporting plants, animals and humans
> - Supporting built structures
> - Returning a yield of agricultural products
> - Bearing groundwater and mineral deposits
> - Providing opportunities for human recreation and aesthetic enjoyment
> - Serving as an archive of natural history
> - As a host medium, serving as a genetic reservoir.
>
> **Waters**
>
> - Serving as a drinking water resource
> - Providing a source of irrigation
> - Supporting aquatic plants and animals
> - Providing a source of process and cooling water for industry
> - Providing a source of renewable energy
> - Providing a means of transport
> - Providing opportunities for human recreation and aesthetic enjoyment.
>
> Some of the soil functions may have been compromised already in respect of most industrial sites and the question arises as to what extent efforts should be made to restore a full range of functions, particularly where these are not considered relevant to the immediate planned development.
>
> It has not been UK practice to seek to return soils to a full multi-functional condition prior to re-use, and this is reflected in the land-use classification of UK trigger concentration values. Similar considerations apply to the development of Statutory Water Quality Objectives where the use of the water is taken into account in the derivation of the value. However, in the application of such guidelines, and during the formulation of site-specific values, it is important to address both the long-term and wider (e.g. groundwater quality) implications of the action and remediation values adopted. Thus in the case of a contaminated site located close to a major water resource, the derivation of site-specific values must take the presence of soluble, or otherwise mobile, hazardous substances into account.

Non-dedicated standards and guidelines can play an important role in data assessment without resource to site-specific estimation of risks by considerably extending the ability of the risk assessor to apply data generated in related fields of health and environmental protection to the assessment of contaminated land.

Site-specific quantified risk assessment involves the detailed evaluation of critical hazards, pathways and targets to establish the nature, magnitude and likelihood of exposures and effects (see Section 6). During the process, the exposure of the target to the contaminant under carefully defined conditions is calculated, and an assessment made of the possible effects. Non-dedicated standards and guidelines providing comparable measures of acceptable exposure are then used to assess whether calculated values are likely to result in adverse effects. Judgements on the acceptability or otherwise of the calculated exposure value and associated risks (and hence the significance of the observed levels of contamination), i.e. risk evaluation, are based on the outcome of the comparison.

The main advantages of quantified risk assessments include :

- the ability to define site-specific action (and remediation) values that are tailor made for the specific risk requirements of the site (i.e. they should be neither over or under protective)

- applicability to any hazard/pathway/target scenario provided sufficient site characterisation, toxicological or other data on effects can be obtained; this avoids the 'list mentality' where only the scenarios covered by generic guidance, rather than those actually identified at the site, are evaluated

- the insight into the behaviour of contaminants gained as a result of the detailed assessment may be beneficial in the subsequent selection and evaluation of alternative remedial strategies.

The main disadvantages of quantified risk assessments include :

- the additional time and effort required compared to an assessment using generic values

- the potential for variations in the interpretation of data depending on the assumptions and uncertainties involved.

Quantified site-specific risk assessment also requires specialist expertise and data handling techniques (e.g. modelling) which may not yet be routinely available in the UK.

An intermediate approach has been developed by the National Rivers Authority in the West Midlands where 'generic' river catchment values have been identified by the authority to assist in the assessment and remediation of contaminated industrial sites on a site-specific basis.

6.3.6 Guidelines vs standards

Guidelines (*cf.* the ICRCL trigger values) rather than standards are the preferred means of deciding the need for investigation and remedial action for contaminated sites in the UK. There are sound technical and scientific reasons for this policy (in addition to any administrative advantages).

Generic standards, as defined here, are statutory limits which must be applied in all cases within the relevant regulatory framework. The application of such binding quality standards to contaminated land problems does not permit site-specific factors, which might require a more restrictive limit to be imposed or permit a relaxation of the standard, to be taken into account. For example, it makes it difficult to allow for possible synergistic effects.

In practice the environment is highly variable, different species exhibit a range of responses to particular concentrations of contaminants, responses vary depending on ambient conditions (e.g. temperature, presence of other contaminants etc.) and other factors, and the combined effects of a number of contaminants may be much more than simply additive. This means the risks presented by contaminated land also vary significantly from site to site. A range of other technical, economic and social factors are also relevant in determining whether and what type of remedial action is actually taken at any particular site.

This suggests generic guidelines which provide a broad level of protection but permit a degree of flexibility in application to take site-specific factors into account (e.g. in relation to the use of the land), are of more practical value when dealing with the reality of most contaminated sites.

6.4 THE USE OF GUIDELINES AND STANDARDS FOR REMEDIATION

6.4.1 Identification of remediation values

Remediation values define the performance objectives to be met by remedial action. Remediation values serve to answer the question 'how clean is clean enough ?' This is a more appropriate question than 'how clean is clean ?' which is better regarded as an expression of the task of establishing background concentration values.

Generic guidelines for remediation have not been set in the UK or by other government bodies although the original Dutch A values (see Appendix 11) were intended to be used in this way. Remediation criteria (guidelines) have been produced in Canada (see Appendix 11).

The general lack of generic remediation values reflects the greater scope, and need, to take site-specific factors into account when developing a technically sound and financially feasible solution to contamination problems. In practice therefore, remediation values are typically set either :

- by adopting or modifying a generic action (e.g. background or threshold) value (see Section 6.3.4)

- by deriving a value based on a site-specific risk assessment.

Where remediation values are set above background concentrations, they should be regarded as minimum standards to be achieved. If the remedial strategy permits a higher standard (i.e. lower concentrations) to be reached without undue additional cost, advantage should be taken of this capability.

In no circumstances should remediation values be regarded as limits to which concentrations can be permitted to rise by addition of more contaminated soil or water. Remediation values are not a licence to contaminate the environment.

6.4.2 Implications for selection of methods

The principles of Integrated Pollution Control (IPC) which embraces the concepts of BATNEEC (Best Available Techniques Not Entailing Excessive Cost), and BPEO (Best Practicable Environmental Option) will apply to many process-based methods used in the treatment of contaminated land. Non-dedicated standards applying to the operation of such plant may influence the selection of remedial methods for particular applications. For example, limits may be imposed on atmospheric emissions and discharges of water arising from treatment plant which are more difficult to achieve than the performance requirements (remedial objectives) adopted for the plant itself.

6.5 DEALING WITH VARIABILITY

A guideline or standard is a single value applicable to a defined situation (e.g. the concentration of cadmium in an allotment soil, the concentration of trichloroethylene in the discharge of a groundwater treatment system). However, sites may occupy many hectares, the distribution of the contaminant may be very uneven, the concentrations of contaminants in untreated and treated media very variable, and the data limited in quantity. The reliability of the analytical data may also, in practice, be uncertain. Thus great care is required when judging observed concentrations of contaminants against guidelines or standards, or applying remediation values to the outcome of any remedial action (see Section 7).

In addition to a consideration of site-specific factors as outlined above, judgements must be made on the basis of:

- an appraisal of the raw data obtained at the site investigation, and attendant, stages with or without the assistance of statistical procedures describing the distribution of the data[7,11]
- a consideration of the reliability of the analytical data in terms of its sensitivity and accuracy[16-18]
- the use of individual spot (not composite or averaged) sample results for assessment purposes
- advance agreement with the relevant regulatory authority on the statistical basis (limits and sampling frequency) of any evaluation of performance in relation to site-specific remediation values.

6.6 UK GUIDELINES AND STANDARDS

6.6.1 Scope of UK guidelines and standards

Both dedicated guidelines, and non-dedicated guidelines and standards are available in the UK. Dedicated guideline values are available for assessment purposes only.

6.6.2 Soils and fills

A limited number of guideline values are currently available in the UK:

- the ICRCL Trigger Concentration Values[5]
- the ICRCL Trigger Concentrations for mine spoil[19]
- Building Research Establishment guidelines on sulphate and acid resistance of concrete in the ground.[20]

Guidance from the Ministry of Agriculture, Fisheries and Food (MAFF) on the protection of soil[21] may be useful in some instances.

None of these guidelines is reproduced here. All undergo change from time to time and it is essential that practitioners always refer to current information.

Available guidelines relate to chemical contaminants. None relate to the soluble fraction of a contaminant in soil which is the most relevant to the protection of water resources. There are no directly applicable guidelines on biological agents, although guidance is available on other potential hazards such as combustibility[22,23]. Formal standards apply to radioactive materials under the Radioactive Substances Act and related regulations.

ICRCL Trigger Concentrations

The ICRCL trigger values, and the rationale for their use, are published in Guidance Note 59/83[5]. More detailed information relating to the derivation of trigger concentration values in respect of former coal carbonisation plant is published in a separate report[13]. The government has announced its intention to extend the range of the guidance[24] and is working on the production of a framework in which generic and site-specific criteria can be derived in a consistent and rational way[25].

The ICRCL trigger values are action guidelines:

- the threshold trigger value **indicates** the concentration above which it may be necessary to carry out additional investigation and/or take some form of remedial action

- the action trigger value **indicates** the concentration above which it is likely that some form of remedial action will be required (possibly following additional investigation).

Depending on the particular circumstances, site-specific values prompting further action may be set anywhere in the 'grey zone' between the two trigger values. Where an action trigger value is not available one must be derived or reference made to other sources of guidance (see below). In some limited circumstances it may be appropriate to set action values below the threshold trigger value.

Each trigger value takes into account only certain hazards. For example, those for nickel and copper relate only to phytotoxicity; they do not take human toxicity (copper) or allergenic reactions (nickel) into account. If potential hazards other than those encompassed by the guidelines are identified site-specific values must be derived.

ICRCL Guidelines for metalliferous mining sites

The ICRCL notes on the restoration and aftercare of metalliferous mining sites for pasture and grazing[19] present 'threshold trigger concentrations' and 'maximum (action trigger) concentrations' (values not to be exceeded for a specified use) for a limited range of elements. Maximum concentrations are provided for grazing livestock and for crop growth (risk of phytotoxic effects). The maximum value is set at a concentration 'above which there is a very high probability of phytotoxic or zootoxic effects which may result in death of stock if the quoted concentrations are *continually exceeded*.' The guidelines take account of the generally limited biological/environmental availability of typical contaminants, e.g. cadmium and arsenic, when present as the chemical species typical of such sites. They also assume high standards of husbandry and recognise that some sub-clinical effects may occur in animals at concentrations between the two values depending on local conditions and management practices. It is important that these guidelines are not applied in applications other than those specified.

BRE Guidance on sulphate and acid resistance of concrete in the ground

Guidance in BRE Digest 363[20] is presented as a series of action guidelines. According to the Digest, if specified concentrations of sulphate or pH levels are exceeded in samples collected from the site, certain precautionary measures, relating to cement type and mix proportions, are deemed necessary.

MAFF Code of Practice for Protection of Soil

The MAFF *Code of Good Agricultural Practice for the Protection of Soil*[21] includes a discussion of the concentrations of a range of inorganic contaminants at which adverse effects on livestock or plants may be observed and the factors which may control uptake. Limiting concentrations in a range of circumstances are suggested. Note that for industrial contamination the code states:

> *It is unlikely that industrial sites, other than old mining areas, will be returned to agriculture. However, if this is planned the ICRCL garden and allotment thresholds should apply, except in the case of grazing on old mine sites where separate guidelines[19] have been prepared.*

The GLC guidelines

During the 1970s the Greater London Council (GLC) produced guidelines for a range of contaminants which classified soils as 'uncontaminated', 'contaminated', 'heavily contaminated' etc. on the basis of measured concentrations[26]. This arbitary classification was based on the GLC's experience of the quality of soils in London and the guidelines were developed to help the Council to make judgements about how to deal with the large number of sites for which it was responsible. They were also used to facilitate decisions on appropriate disposal locations for contaminated excavation arisings. They reflect frequency of occurrence of contamination levels on sites known to be contaminated and do not relate directly to levels of risk. The limiting concentration for 'uncontaminated' soils was set in a number of instances (e.g. lead) well above natural background concentrations. Once the ICRCL trigger values became available in 1979, the GLC switched to these for use in hazard assessments and decisions about the need for remedial works. Therefore, the GLC guidelines **should not** be used as a basis for assessing the risks associated with contaminated land or for assessing potential hazards to workers (even though they are reproduced in a recent Health and Safety Executive publication[27]).

6.6.3 Surface and groundwaters

Dedicated surface and groundwater guidelines have not yet been produced in the UK.

Reference can be made to non-dedicated values, such as those contained in the various EC Directives on water quality[12] (including drinking water); guidance[28] provided by the Department of the Environment on the monitoring and surrender of waste management licences may also be of relevance. In due course, it will be possible to make comparison with the Water Quality Standards objectives set for surface waters which are contained in the Surface Waters (Rivers Ecosystem) (Classification) Regulations 1994[29, 30]. Objectives will be set for other types of controlled waters in due course. Dangerous substances which might be discharged to the environment are commonly categorised as black, grey or red list substances under EC or UK legislation or regulations (see Box 6.6).

EC Directives

Relevant values are presented in the form of guideline and mandatory concentrations of a range of substances in waters intended for specific purposes, including drinking and bathing water, and water intended to support fish life and shellfish. Guideline values indicate that quality of water which Member States are required to endeavour to meet. Member States are legally obliged to meet the standard indicated by mandatory limits, since these are intended to protect public health and other similarly fundamental requirements.

Approximately 50 parameters are prescribed under the Directive on the quality of surface water intended for the abstraction of drinking water[31-33], including individual elements and compounds; groups of compounds (e.g. dissolved or emulsified hydrocarbons); indicator parameters (e.g. biological and chemical oxygen demand) and microbiological agents. A complementary Directive specifying reference measurement (sampling and analysis) methods was adopted by the Council of Ministers in October 1979[36]

Other Directives have a similar structure and intent, although the presentation of the standards may differ slightly. The Directive on the quality of surface waters needing protection or improvement in order to support fish life[37] provides guideline and mandatory standards in relation to both salmonoid and cyprinid waters. Specific compliance conditions (percentage of measurements) apply to certain test parameters, such as dissolved oxygen. In other cases, limits apply in respect of average readings.

> **Box 6.6** *Black, grey and red list substances*[35]
>
> **List I and II substances ('Black' and 'Grey' Lists)**
>
> Directive 76/464/EEC[31] created a framework for the elimination or reduction of pollution of inland, coastal and territorial waters by dangerous substances. Control is achieved through various 'daughter' Directives which lay down standards for particular substances. Substances are divided into two categories:
>
> (I) those considered to be most harmful – pollution from these must be eliminated, and
>
> (II) less harmful substances – pollution from these must be reduced.
>
> Lists I and II are commonly known as the black and grey lists, respectively. In broad terms, List I substances are controlled at EC level, with standards for discharges or for environmental quality which apply across the EC, while List II substances are controlled at national level.
>
> Procedures and policies for the implementation of 76/464/EEC in England and Wales are described in a DOE Circular[32].
>
> Discharges of List I substances are controlled through:
>
> - a limit value (LV), a uniform emission limit to be complied with irrespective of the size or number of plants or the nature of the receiving water (favoured by most countries), or
>
> - the setting of environmental quality standards (EQSs) which are concentrations not to be exceeded at particular locations in the receiving water (the EQS is set with the aim of protecting all aquatic life) – the option generally but not exclusively used in the UK.
>
> For List II substances member states must develop programmes which ensure that appropriate quality objectives are met in receiving waters.
>
> The Groundwater Directive (80/68/EEC)[33] also refers to the List I and II substances. It requires that member states take specific measures to prevent List I substances from entering groundwater, and to restrict the entry of List II substances into groundwater. The same criteria apply whether the aquifer is in use or not.
>
> **Red list substances.**
>
> In 1988 DOE proposed[34] a new policy for control of the most dangerous substances to water combining the use of Limit Values and Environmental Quality Standards. The proposals were intended to provide the basis for the UK to implement the commitments made at the first and second North Sea conferences. The first list of substances subject to these stringent controls was published in 1989[35]. The 'red list' is essentially a sub-set taken from the EC priority candidate black list. A first priority candidate red list was published at the same time. A consultation document containing proposed EQSs for those substances on the first red list for which EQSs had not already been set at EC level was issued in 1991.
>
> **Further information**
>
> For a summary account of EC and UK provisions for the release of dangerous substances to water and their interactions see Reference 12.

As with all standards or criteria borrowed for the purpose of assessing the analytical output of contaminated site investigations, care should be exercised in the selection and application of water quality standards, particularly when standards are exceeded.

Guidance on the surrender of waste management licences

Waste Management Paper 26A[28] provides default criteria for leachate which might enter groundwater for use at 'sites for which locally based criteria have not been developed'. It is assumed that any leachate entering groundwater will be diluted by mixing, by a dilution factor of at least ten. To protect potable water sources, default water criteria are based on 10 × Maximum Allowable Concentrations or Guide Levels in the Drinking Water Directive, except Total Organic Carbon which is based on the background level of 1mg/litre carbon in 'clean water.' In respect of groundwater, it is stated 'normally no change in groundwater quality should be observed between monitoring points upgradient and downgradient from the site' and similarly for surface waters.

6.6.4 Ambient air

Criteria relating to acceptable concentrations of contaminants in the air are relevant from the point of view of worker health and safety, and in relation to the protection of public health and the prevention of nuisance.

Work place health and safety is regulated under various statutes, including for example, the COSHH Regulations and contaminant specific regulations such as those for lead, asbestos and radioactive materials. Control is exercised through criteria such as Occupational Exposure Standards (OESs) and Maximum Exposure Limits (MELs) – see Box 6.7. The procedures for application of these criteria are well established and documented elsewhere (see Volume XII) and are not discussed further here.

Box 6.7 *Occupational exposure limits*

There are two types of occupational exposure limit defined in Regulation 2 of the COSHH Regulations and applied in Regulation 7.

Occupational Exposure Standards (OESs) are set at a level at which there is no indication of risk to health.

Maximum exposure limits (MELs) take socio-economic factors into account and are set at a level at which a residual risk may exist.

OESs and MELs are set on the recommendation of the Advisory Committee on Toxic Substances (ACTS) following assessment by the Working Group on the Assessment of Toxic Chemicals (WATCH), of toxicological, epidemiological and other data. The criteria applied to decide the type of limit which is appropriate and the concentration at which it should be set are set out in HSE guidance note EH40[38] which is revised annually. This document lists current occupational exposure limits and provides guidance on their application.

With the exception of ambient air quality standards for sulphur dioxide, nitrogen oxides, ozone, particulates and lead (see Volume XII) there is a general lack of air quality guidelines, criteria or standards. In many cases, it will be necessary to devise site-specific criteria through a detailed risk analysis, or by adaption of non-dedicated values (such as OESs). Guidance produced by HMIP on the emission limits applying to prescribed processes (under the Environmental Protection Act 1990) may be helpful in some applications indicating the types and ranges of concentrations considered acceptable from point sources.[39] However, the rationale for and technical basis of such guidance needs to be understood as they are, at least in part, technology related.

Airborne materials which may have to be addressed in relation to the environmental impact of contaminated sites include dusts, 'bulk gases' such as methane and carbon dioxide, and trace gases, such as volatile organic compounds. Both the toxicity and amenity (e.g. odours) aspects of atmospheric emissions require consideration.

OESs are based on the protection of reasonably healthy adults of working age under relatively controlled conditions during a normal working week (40 hours). A control limit for a wider population must take into account the potential exposure of more sensitive sections of the population, such as the elderly, the young, infirm and pregnant; longer exposure times (up to seven days (168 hours) a week); and the near impossibility of organising and implementing any systematic health surveillance. It may be necessary also to take into account other airborne sources, e.g. indoor air pollution, which may also be impacting on the target.

It is customary to work on the basis of some small fraction of the OES, usually no more than one fortieth, and sometimes, depending on the chemical, a lesser fraction (even as low as 1/420 or 1/1000). The basis for this type of value is that exposure may occur over periods up to 4.2 times longer than occupational exposures, together with the application of safety (uncertainty) factors of between 10 and 100. Risk-based derivation of ambient air quality criteria may produce lower values than this rule-of-thumb approach.

The presence of detectable or identifiable (in terms of the compound(s) present) odour can provide a very direct measure of acceptability in amenity terms, but various factors have to be taken into account when deciding whether odour is, or is not, present. The apparent intensity of an odour may vary with weather conditions, and individuals vary significantly in their ability to detect odours. Individuals can also become conditioned to the presence of an odour over time and may therefore be unaware of its presence.

There is no consistent relationship between odour thresholds and toxicity. Some compounds may be toxic at concentrations below the detectable odour threshold; the opposite is true for other substances. Therefore the absence of detectable odour may not be an adequate guarantee of the absence of a health risk. Information on odour threshold and identification values for a range of substances found in association with contaminated sites in the US are available (see for example reference 40).

Using a combination of values derived as fractions of OESs and similar values, together with a consideration of odours, site-specific action values for ambient air quality can be defined. Appropriate action levels might be:

- a lower concentration limit (say 1/100 of the relevant OES) which if exceeded triggers further investigation
- a higher concentration limit (say 1/40) of the OES above which action is required to control the source of the airborne contaminant.

6.5.5 Below ground gases

Practitioners should be aware of the possible presence, and implications of in-ground gases such as methane, carbon dioxide, radon and oxygen depletion, particularly in relation to potential health and safety, nuisance and more serious off-site effects (e.g. explosion). A detailed consideration of these contaminants is beyond the scope of this report. Waste Management Paper 26A on landfill completion[28] provides guidance on the concentrations and emission rates of methane and carbon dioxide which would enable a site to 'be regarded as stabilised.' Other guidance on acceptable concentrations in relation to existing and planned development may also be relevant[41-44].

REFERENCES

1. INTERNATIONAL ORGANISATION FOR STANDARDISATION. *Soil Quality – Vocabulary – Part 1: Terms and definitions relating to soil protection and pollution* (Available from the British Standards Institution, London)

2. SMITH, M. A. Standards for the redevelopment of contaminated land. In: *Proceedings of a Conference on the Reclamation of Contaminated Land*. Society of Chemical Industry (London), 1980, pp B1/1-B1/16.

3. SMITH, M. A. Investigation of contaminated sites: standardisation for investigation, testing and analysis. In: *Proceedings Conference on Contaminated Land Site Investigation (London 1993)*. IBC Technical Services (London) 1993.

4. *Proceedings of a Conference on Developing Clean-up Standards for Contaminated Soil, Sediment and Groundwater: How Clean is Clean ?* Water and Environment Federation (Alexandria, Virginia), 1993

5. DEPARTMENT OF THE ENVIRONMENT. *Guidance on the Assessment and Redevelopment of Contaminated Land*. ICRCL Guidance Note 59/83 (2nd Edition). DOE (London), 1987

6. ROYAL COMMISSION ON ENVIRONMENTAL POLLUTION. *Tackling pollution – experience and prospects*. Tenth Report. HMSO (London), 1984

7. SMITH, M. A. Data analysis and interpretation. In: *Recycling derelict land*. G. Fleming (ed.) Thomas Telford (London), 1991, pp 88-144

8. ALLOWAY, B. J. *Heavy Metals in Soils*. B. Alloway (ed.). Blackie (Glasgow), 1990

9. McGRATH, S. P. and LOVELAND, P. J. *The Soil Geochemical Atlas of England and Wales*. Blackie Academic Publishers (Glasgow) 1992.

10. SINCLAIR, A. J. Statistical analysis of trace element data. In: *Applied Science Trace Elements*. B. E. Davies (ed.)Wiley (Chichester), 1980, pp 131-153

11. HEWITT, C. N. *Methods of Environmental Data Analysis*. C. N. Hewitt (ed.) Elsevier (London), 1992

12. ANON. *Dangerous Substances in Water: A Practical Approach*. Environmental Data Services (London), 1992

13. DEPARTMENT OF THE ENVIRONMENT. *Problems arising from the redevelopment of gas works and similar sites*. DOE (London), 1987

14. SMITH, M.A. Experiences of the development and application of guidelines for contaminated sites in the United Kingdom. In: *Proceedings of a Conference on Developing Clean-up Standards for Contaminated Soil, Sediment and Groundwater: How Clean is Clean ?* Water and Environment Federation (Alexandria, Virginia), 1993, pp 195-204

15. SMITH, M. A. Identification, investigation and assessment of contaminated land. *Water and Environmental Management*, 1991, **5** (6), 616-623

16. HAMILTON, E. I. Analysis for Trace Elements I: Sample treatment and laboratory quality control. In: *Applied Science Trace Elements*. B.E. Davies (ed.)Wiley (Chichester), 1980, pp 21-68

17. HAMILTON, E. I. Analysis for Trace Elements II: Instrumental analysis. In: *Applied Science Trace Elements*. B.E. Davies (ed.)Wiley (Chichester), 1980, pp 69-130

18. LORD, D. W. Appropriate site investigations. In: *Reclaiming Contaminated Land*. T. Cairney (ed.)Blackie (Glasgow), 1987, pp 62-113

19. DEPARTMENT OF THE ENVIRONMENT. *Notes on the Restoration and Aftercare of Metalliferous Mining Sites for Pasture and Grazing*. ICRCL Guidance Note 70/90. DOE (London), 1990

20. BUILDING RESEARCH ESTABLISHMENT. *Sulphate and acid resistance of concrete in the ground*. BRE Digest 363. BRE (Garston), 1991

21. MINISTRY OF AGRICULTURE, FISHERIES AND FOOD/ WELSH OFFICE AGRICULTURE DEPARTMENT. *Code of Good Agricultural Practice for the Protection of Soil*. HMSO (London), 1993

22. DEPARTMENT OF THE ENVIRONMENT. *Notes on the fire hazards of contaminated land*. ICRCL Guidance Note 61/84 (2nd edition). DOE (London), 1986

23. LUCAS, R.H. and CAIRNEY, T. Reclaiming potentially combustible sites. In: *Contaminated Land − Problems and Solutions*. T. Cairney (ed.) Blackie (London), 1993, pp 141-159

24. DEPARTMENT OF THE ENVIRONMENT. *The Government's Response to the First Report from the House of Commons Select Committee on the Environment: Contaminated Land*. Cm 1161. HMSO (London), 1990

25. FERGUSON, C. and DENNER, J. Soil remediation guidelines in the United Kingdom: A new risk based approach. In: *Proceedings of a Conference on Developing Clean-up Standards for Contaminated Soil, Sediment and Groundwater: How Clean is Clean ?*. Water and Environment Federation (Alexandria, Virginia), 1993, pp 205-212

26. KELLY, R. T. Site investigation and material problems. In: *Proceedings of a Conference on the Reclamation of Contaminated Land*. Society of Chemical Industry (London), 1980, pp B2/1-B2/14

27. HEALTH AND SAFETY EXECUTIVE. *Protection of workers and the general public during the development of contaminated land*. HS(G)66. HMSO (London), 1991

28. DEPARTMENT OF THE ENVIRONMENT/ SCOTTISH OFFICE/ WELSH OFFICE. *Waste Management Paper 26A: Landfill Completion*. HMSO (London), 1992

29. DEPARTMENT OF THE ENVIRONMENT. *River Quality: the Government's Proposals*. A Consultation Paper. DOE (London), 1992

30. *SI 1994 No. 1057, The Surface Waters (River Ecosytem) (Classification) Regulations 1994*. HMSO (London) 1994

31. *Directive on pollution caused by certain substances discharged into the aquatic environment of the Community*. Directive 76/464/EEC. OJ L129, May, 1976

32. DEPARTMENT OF THE ENVIRONMENT. *Water and the Environment: the implementation of the European Community Directives on pollution caused by certain dangerous substances discharges into the aquatic environment.* Circular 7/89. DOE (London), 1989

33. DEPARTMENT OF THE ENVIRONMENT. *EC Directive on protection of groundwater against pollution by certain dangerous substances (80/68/EEC): Classification of listed substances.* Circular 20/90. DOE (London), 1990

34. DEPARTMENT OF THE ENVIRONMENT. *Inputs of Dangerous Substances to Water: Proposals for a Unified System of Control.* DOE (London) 1988

35. ANON. *Dangerous Substances in Water: A Practical Approach.* Environmental Data Services (London), 1992

36. *Directive concerning the methods of measurement and frequency of sampling and analysis of surface water intended for the abstraction of drinking water in Member States.* Directive 79/869/EEC. OJ L271, October, 1979

37. *Directive on the quality of fresh waters needing protection or improvement in order to support fish life.* Directive 78/659/EEC. OJ L222, August, 1978

38. HEALTH AND SAFETY EXECUTIVE. *EH 40/94 Occupational Exposure Limits 1994.* HMSO (London) 1994

39. HER MAJESTY'S INSPECTORATE OF POLLUTION. *Chief Inspector's Guidance to Inspectors, Environmental Protection Act, 1990, Process Guidance Notes (various titles).* HMSO (London)

40. SMITH, M.A. *Contaminated Land: Reclamation and Treatment.* Appendix K. Plenum (London), 1985, pp 407-417

41. ANON. *Building Regulation 1991: Approved Document C: Site Preparation and Resistance to Moisture.* HMSO (London), 1992

42. BUILDING RESEARCH ESTABLISHMENT. *Construction of new buildings on gas contaminated ground.* BRE (Garston), 1991

43. BUILDING RESEARCH ESTABLISHMENT. *Radon: Guidance on protective measures for new dwellings.* BRE (Garston), 1991

44. DEPARTMENT OF THE ENVIRONMENT. *Landfill Gas.* WMP No. 27. HMSO (London), 1991

Further reading

HARRIS, R. and THOMAS, C. A. Contaminated land and water quality standards. In: *Proc. Conference on Contaminated Land: Policy, Risk Managment and Technology, London 1994.* IBC Technical Services (London) 1994

SOCIETY FOR ENVIRONMENTAL GEOCHEMISTRY AND HEALTH. WIXSON, B. G. and DAVIES. B. E. *Lead in soil: recommended guidelines.* Science Reviews (Northwood) 1993

7 Investigation for compliance and performance

7.1 SCOPE

Investigation (or monitoring) for compliance and performance will be necessary during and following remedial action to determine whether:

1. A removal action (e.g. excavation) has been successfully completed.

2. An ex-situ treatment process is achieving the required performance in terms of reducing contamination levels in treated water or soil.

3. In-situ soil or groundwater remediation has been successful or is proceeding as expected.

4. A process to treat groundwater on an ex-situ basis has been successfully completed (as measured by the quality of the treated groundwater) or is having the desired result (in terms of reducing in-situ contamination levels).

5. Containment measures are working as planned.

This Section is concerned mainly with these five applications. Other monitoring/measuring activities typically relating to the operational phase of remediation are mentioned where appropriate, including:

- monitoring process feedstocks — in support of process optimisation and to assist in meeting performance and regulatory requirements

- monitoring emissions and waste streams for both (regulatory) compliance and process control purposes

- monitoring around an in-situ treatment zone to detect possible migration/escape of contaminants or treatment agents.

It may also be desirable in some cases to evaluate the performance of a remedial process in a broader sense (see Box 7.1), for example:

- as part of a demonstration of the technology

- to evaluate how a remedial system (e.g. containment) works in practice so as to inform future design and execution.

In all cases it will be necessary to decide:

- the sampling and analytical strategies to be used for demonstrating compliance with pre-determined remedial objectives (e.g. permitted residual concentrations)

- the statistical basis on which compliance is to be judged

- quality assurance and control procedures relating to the compliance testing

- the response required to different levels of non-compliance.

Precise requirements will depend on the application and in particular whether:

- the intention is the complete destruction or removal of contaminants
- the contaminants are not removed or destroyed (e.g. solidification or containment)
- treatment continues over an extended period of time (possibly decades in the case of some pump-and-treat operations).

Performance and compliance assessment requirements for a range of different remedial methods are discussed below. The general philosophy regarding the long-term effectiveness of remedial methods and the application of monitoring have been discussed by Stief[1] and by Cairney[2]. The former lists important points and questions to be addressed in relation to different remedial methods.

Box 7.1 *Monitoring versus evaluation*

The distinction between monitoring and evaluation made here follows that made by Stief[1]:

- *Monitoring* is intended to show whether the remedial measures used are continuing to work as required: it should be designed to give an indication of actual or, preferably, possible failure at some time in the future.
- *Evaluation* is intended to determine the quality of execution and how the remedial measures are working to provide data that will improve future design and execution i.e. it is essentially a process of research.

7.2 EXCAVATION

Excavation may be a precursor to:

- disposal off-site
- disposal on-site
- treatment in a process-based system.

Excavated material will usually be replaced by imported 'clean' fill or by the clean product of a remedial process.

In each case, the primary concern is to demonstrate satisfactory removal of material containing contaminants above a specified level. It may also be necessary to confirm that material transferred for disposal complies with pre-determined acceptance criteria (for the proposed disposal route), or that its composition and/or physical properties are within the range that can be dealt with by a selected remedial process. In addition, it will often be necessary to confirm that any groundwater (which may in some cases require on-site treatment prior to discharge) complies with any discharge consents.

Experience has shown that it is also essential to monitor the quality of any imported fill materials before placement.

The procedures for validation of an excavation operation are discussed in detail in Volume V; the guidance given on sampling strategies in Section 4 should also be noted. The requirements of the three main post-excavation options are summarised in Table 7.1.

Table 7.1 *Compliance/performance testing for excavation based operations*

Excavation strategy	Typical compliance/performance requirements
Common requirements	Testing status of material remaining in place on the site following excavation (see Volume V)
	Testing of contaminated groundwater etc. for compliance with discharge consent(s)
	Testing of liquid wastes (e.g. floating phases, tank contents) for compliance with requirements of chosen disposal route
	Testing of quality of imported or replacement fill material including both chemical and physical properties
Excavation for disposal	Testing of excavated material for compliance with agreed composition limits (and sometimes physical properties) for each disposal route employed on agreed basis (e.g. number of samples/volume excavated or shipped)
Excavation for on-site disposal	Testing of excavated material for compliance with agreed composition etc. for materials to be deposited on-site
	Compliance testing of material not suitable for on-site disposal for alternative disposal route
	Testing of placed materials for compliance with any physical parameters e.g. compaction, permeability etc.
	Testing of containment system and imported cover materials etc.
Feed to soil treatment plant	Testing of material to confirm that its chemical and physical properties are within the range that can be handled by the chosen treatment train
	Compliance testing of materials following alternative treatment/disposal route

7.3 PERFORMANCE OF EX-SITU PROCESS-BASED METHODS

7.3.1 General considerations

An ex-situ treatment process involves feeding contaminated material (soil, sediment or groundwater) through a process to obtain a 'clean' product (see Volumes VII & VIII for process descriptions). Usually numerical remedial objectives (see Section 6) will be adopted as the targets to be achieved by the process. These are generally expressed as concentrations of specified chemical species but may also refer to the presence of biological agents, or residual combustible or putrescible matter, or relate to satisfactory performance in tests for toxicity. To confirm that a treatment process is actually meeting set criteria the basis for the performance assessment should be established before treatment begins. Consideration must be given to:

- where in the process train to sample
- size and nature of samples
- frequency of sampling
- analytical strategy
- the statistical basis for assessing compliance
- quality assurance and quality control requirements.

Note that compliance cannot simply be assumed by establishing that an unwanted chemical has disappeared from the treated product:

- the analytical strategy must also cover the eventuality that it has been converted into some other (possibly more toxic) form

- the analytical strategy (and remedial objectives adopted) must take account of the possible presence of process residuals (chemical treatment agents, leachable nutrients etc.)

- the monitoring strategy must be designed to detect fugitive losses (e.g. of volatile substances)

- analytical methods must be selected taking into account possible changes in the 'availability' of particular contaminants following treatment (for example, an inorganic contaminant might be changed by thermal or chemical treatment into a form less soluble in the chosen extractant — see Section 3.6.3)

- it may be necessary to monitor a solid product or treated soil in terms of its physical characteristics (e.g. suitability for use as fill) or its life-sustaining capabilities (nutrient status, organic matter content etc.) if it is to be used as top-soil.

Therefore in addition to determining that the (soil or water) product complies with predetermined standards in terms of the presence of specific (hazardous) substances it may be necessary to:

- establish a mass balance for all components of the feed (e.g. solids and water, contaminants, treatment agents)

- sample at intermediate points in the process train for the presence of unwanted by-products

- carry out a range of ecotoxicological testing.

The statistical basis for deciding whether there is compliance must be predetermined. For example, it may be decided that:

- no test result on any sample should exceed the specified remediation value, or

- some fixed proportion (e.g. 1%) of results may exceed the agreed value.

In the second case it will usually be necessary to set a characteristic value that must not be exceeded (for example) 99% of the time (depending on circumstances 95% might be set). To do this:

- sufficient data must be collected during the early stages of operation so that the 'normal' statistical variations associated with a 'uniform' feedstock can be determined, or

- a decision must be made in advance about the amount of variation permitted, e.g. by specifying acceptable mean and 99 percentile values from which a maximum permitted standard deviation can be calculated (see Box 7.3). The latter needs to be done with care so that one outlier does not exert an undue influence on the acceptability of the treated material (see Box 7.2).

Remediation targets are sometimes set on the basis that, for example, 95% of all or specified contaminants should be removed, and the success of a treatment method is often expressed in such terms. Since this is a relative measure, and 5% of the starting concentration may still be a hazardous concentration, such targets are not very helpful. In addition, they may be unnecessarily stringent if concentrations are initially low and 50% removal would result in acceptable concentrations being achieved. Clear, numerical, targets/objectives set in a statistical framework are therefore preferable.

> **Box 7.2** *Assessment of the effectiveness of a thermal treatment process*
>
> Yland and Soczo[13] have described a study in which the effectiveness of thermal treatment in removing polyaromatic hydrocarbons (PAHs) and other contaminants from a sandy soil and a clayey soil was studied. The standard procedure[3] employed in the Netherlands at the time of the study to assess the effectiveness of remediation technologies involved setting limits for mean concentrations and standard deviation *(Se)* such that:
>
> $$Se = \frac{(y-x)}{2.39}$$ where: x = mean value
> y = 99 percentile
>
> A carefully designed statistical sampling programme was applied to both input and output soils (see Box 7.3). Removal rates in excess of 95% were obtained for both soils but the clayey soil did not meet the statistically based requirements for all parameters. Neither the mean nor the 99 percentile were met for PAHs. Although the mean concentration of mineral oil (initial mean concentration 2602 mg/kg) was far below the target mean (10 vs 100 mg/kg – 99.5% removal) and all values were below the permitted 99 percentile (200 mg/kg) the requirement set for the standard deviation was not met. This was because although only one value was just above the required mean (100 mg/kg), eighteen were far below it. It was concluded that some modification to the prescribed requirements of the standard methodology was needed.
>
> This study clearly demonstrates the need to think carefully about the way that compliance requirements are set and to make provision for their formal modification once operating data become available.

Procedures for monitoring processes are well established in the chemical engineering, minerals processing and related industries. In the Netherlands, specific procedures for monitoring soil treatment systems have been devised[3] (see Box 7.2 and Smith[4]). In Canada a protocol has been prepared for monitoring bioprocesses used to treat contaminated soil[11]. The Technology Evaluation Reports produced under the US Environmental Protection Agency's SITE (Superfund Innovative Technology Evaluation) program provide useful examples of process monitoring under demonstration conditions (see Section 7.7) for a range of different processes (e.g. references 5-10). The approach adopted for a number of the projects included in the NATO/CCMS (Committee on the Challenges of Modern Society) pilot study on demonstrations of innovative technologies for soil and groundwater clean-up[12] have been described by Smith[4].

Variations in the composition of the feedstock will, in most cases, influence the composition of treated soil or groundwater. Monitoring of the feedstock is therefore be essential to proper control of the process. As discussed elsewhere in this Report, when soils and similar materials, or highly contaminated liquid phases are to be treated, an homogenisation process will usually be required in advance of the main treatment train.

7.3.2 Stabilisation/solidification processes

Stabilisation/solidification processes present particular problems for monitoring and testing. Appropriate test methods for use in treatability studies and to determine effectiveness in practice are the subject of considerable debate.

> **Box 7.3** *Sampling and assessment procedures for the case study described in Box 7.2*[13]
>
> **SAMPLING PROGRAMME:**
>
> **Feed to process:**
> Before treatment began the average composition of both lots of soil were determined by preparation and analysis of three 'mixed' samples per lot.
>
> During processing 20 samples were taken at regular intervals from the feed, 10 of which were selected at random for analysis.
>
> **Processed soil:**
> 400-600 tonnes of each soil were processed at about 40 tonnes/hr. A total of 60 one kilogramme samples of processed soil were taken at 9 minute intervals over a 9 hour period (1 sample/6.7-10 tonnes).
>
> **Testing procedure for processed soil:**
> The standard procedure employed required a multi-stage analytical programme: the first 20 samples are selected at random; depending on the results a further 20 samples are analysed, and finally the remaining 20 samples may be analysed. The calculated mean value and its standard deviation are examined. The maximum permissable value for the standard deviation *(Se)* is determined on the basis of the requirements set for the residual concentration:
>
> $$Se = \frac{(y-x)}{2.39}$$ where: x = mean value and y = 99th percentile.
>
> For the first stage of the testing procedure, the soil would be passed if:
>
> mean of 20 samples $m_{20} \leq 0.8x$
>
> standard deviation of 20 samples $S_{20} \leq 0.5Se$
>
> For 40 samples the requirements were:
>
> $m_{40} \leq 0.9x$
>
> $S_{40} \leq 0.65Se$
>
> For 60 samples the requirements were:
>
> $m_{60} \leq x$
>
> $S_{60} \leq Se$

Sampling and testing are likely to be required to determine compliance with quality control and assurance requirements at, or shortly after treatment, for example:

- to confirm that the correct mix (treatment agent(s)/contaminated material) have been used
- to confirm that specified limits for leachability have been met
- to confirm that specified physical properties (e.g. setting time, early strength, permeability) have been achieved.

Monitoring may also be necessary to confirm that longer term performance targets (e.g. for leachability, strength, permeability) are being met. Longer term compliance monitoring may require a combination of tests including laboratory specimens stored under controlled conditions, in-situ testing and testing of samples taken from the deposited material.

It is preferable to use readily available standard tests to determine leachability and they may be specified by the regulatory authorities. However, because of the range of solidification and stabilisation processes available, it is important to establish that the testing regime is appropriate to the issues of importance[14,15,16]. An investigation of test methods for solidified wastes was carried out jointly for the US Environmental Protection Agency and Environment Canada in 1991[14].

Factors to be taken into account include:

- the composition of the leach solution (e.g. should it reflect acid rain, landfill leachate or actual groundwater conditions?)

- the buffering and reaction capacities of the material being tested (many systems are alkaline and will readily neutralise a small volume of acid)

- the physical state of the sample to be tested (e.g. finely ground, cast specimen or core exposed on all surfaces, a specimen through which the test solution permeates under hydraulic pressure)

- the volume ratio of test solution:solid to be leached

- whether the test specimen and solution are held static relative to one another (e.g. a solid specimen resting or hanging in a solution — an equilibrium condition will develop possibly involving considerable composition gradients) or not (e.g. solution around solid specimen stirred, whole system shaken)

- the manner of leaching (single extraction, repeated extractions, continuous extraction)

- the assumed mechanisms of leaching or reaction (e.g. diffusion controlled, shrinking-core model[16]).

The selection of appropriate physical tests will also require careful attention.

For both leaching and physical tests the following should be borne in mind:

- in systems based on Portland cement and other cementitious materials it is important to distinguish between 'setting' properties and subsequent strength development

- the importance of heat (usually generated during hydration) to the rate of strength development and to longer-term properties of cementitious systems

- how curing conditions (temperature, moisture, exposure to air at critical stages) can profoundly affect the development of properties

- that properties (e.g. strength, reduction in permeability) may continue to develop over a prolonged period (12 months or more)

- that degradation mechanisms (e.g. chemical attack, freeze-thaw damage) are often slow and may take ten or twenty years to become manifest under field conditions

- that agitation will drive off volatile organic compounds (VOCs) and some semi-volatiles (especially if temperatures rise during processing).

Typically, the treatment product may:

- initially take the form of a slurry which can be pumped or transported to the disposal area, or

- be a relatively low moisture content material which can be placed using standard engineering techniques.

In the first case, it may be necessary to set requirements for the amount of 'bleeding' (segregation of free water during standing) permitted (preferably a very small amount). In the latter case, engineering tests will be required to ensure that prescribed compaction and permeabilities have been achieved.

As placement of either type of material is unlikely to be continuous, consideration must also be given to how joints between successive placements are to be formed, and how compliance with requirements for these is to be determined.

It may sometimes be necessary to protect the stabilised material to aid curing or to avoid damaging effects from adverse weather conditions (e.g. frost, rain, high temperatures).

An important aspect of situations where solidified or stabilised materials are to be placed in the ground is how non-compliance in any of the agreed tests is to be dealt with. Many of the tests employed take several hours to perform, others may take days; yet the time during which the material can be handled for placement may be no more than an hour or so. Thus, it is essential to ensure that non-compliant material can be recovered and safely disposed of or re-treated.

Because of the uncertainties surrounding the long-term performance of some stabilisation/solidification systems, long-term monitoring of disposal areas will be required in many instances (whether or not a contained landfill has been established — if containment measures have been employed their long-term effectiveness will have to be monitored). This may take the form of monitoring around the periphery, monitoring of wells installed at the time of placement within the deposited material, or periodic sampling of the bulk material, e.g. by coring. Coring carries some risk, e.g. of damage to any liner if drilling goes too deep, increased infiltration if the hole is not subsequently properly sealed, release of vapours. The sampling must be representative and the analytical methods employed consistent between sampling episodes.

7.4 REMEDIATION OF GROUNDWATER THROUGH EX-SITU TREATMENT (Pump and treat)

Where groundwater is extracted for treatment on an ex-situ basis, confirmation of effectiveness will usually involve a combination of:

- monitoring the composition of the extracted water (the feed/influent to the treatment plant)
- in-situ monitoring.

Monitoring the performance of the process is governed by the considerations outlined in Section 7.3.

Particular difficulties arise when trying to establish that an ex-situ groundwater treatment has been successful in removing contaminants from the ground (similar difficulties also occur with in-situ treatment — see below). Experience has shown that because it is difficult to remove or contact water and contaminants located in the fine pore structure, concentrations of contaminants in the groundwater may initially drop, but then rise again as further contaminants are released from the fine pores (see Figures 7.1 and 7.2). These difficulties can be partly overcome by cyclic pumping (see Figure 7.3). Similar effects occur where there is a dense non-aqueous phase lying at the base of an aquifer. Changes in the water table brought about by pumping may also result in the movement of lighter than water non-aqueous phases below the permanent water table thus increasing the chances that they will contaminate the groundwater when pumping is halted (see Figure 7.4, and Volume VI).

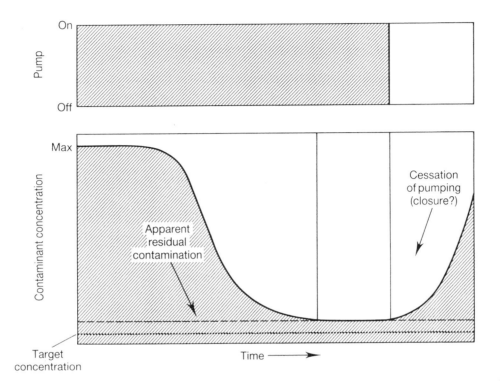

Figure 7.1 *Contamination increases after remediation stops: contaminant concentrations in groundwater may rebound when pump-and-treat operations cease because of residual contaminants trapped in capillaries (see Figure 7.2).*

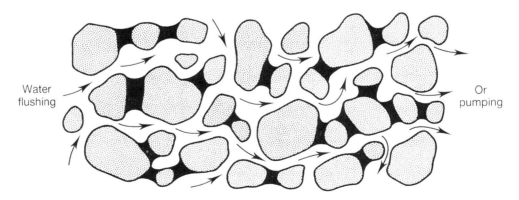

Figure 7.2 *Limitation of effectiveness of flushing action due to contaminants trapped by capillary action*

Testing procedures must address:

- sampling locations
- number of samples
- frequency of sampling
- period of sampling
- analytical strategy
- quality assurance and control requirements
- the statistical basis for compliance.

The statistical basis for compliance should include both remedial objectives and the time period over which they are to be met in order to confirm that treatment is complete.

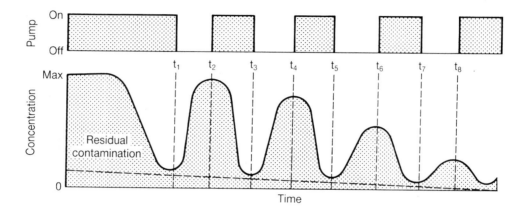

Figure 7.3 *Reduction of residual contaminant mass by pulsed pumping*

The quality of water extracted from the site itself is unlikely to be sufficient on its own to confirm effectiveness because it may derive from preferential paths within the aquifer. It also provides no information on the further reaches of the plume; these are not directly affected by treatment but they should become progressively 'cleaner' as source concentrations diminish. A number of permanent sampling wells are likely to be required and it may also be desirable from time to time (and certainly when treatment has apparently been 'completed') to extract cores to determine the extent to which contaminants have been physically or chemically adsorbed onto the host medium.

In practice, the achievement of (say) 'drinking water standards' at specified locations away from the site might be taken to be the (first) measure of success: although the groundwater close to the source remains contaminated it need no longer be regarded as a (major) 'problem.'

7.5 PERFORMANCE OF IN-SITU PROCESSES

7.5.1 General considerations

Although the discussion below has been separated into soils and groundwater, in practice many in-situ methods of remediation affect both media, or they will be separately affected by parallel treatment processes, e.g. for example combined soil vapour extraction, water extraction and microbial treatment can together treat the unsaturated zone, floating layers of non-aqueous liquid, and contaminated groundwater. Descriptions of in-situ methods are provided in Volume IX.

In practice, a range of performance measurements will be required during and after such combined treatment. Mass balances may need to be established. Testing should take place whilst remedial works are in progress, and on 'completion' of the works. The former can be done by monitoring various process streams and using fixed monitoring points: the latter is more likely to require the installation of additional monitoring or sampling points, or even the full re-investigation of the site. Stief[1] discusses a number of issues which must be addressed when contaminants are not destroyed or removed (as in stabilisation/solidification processes, or surface amendment with lime to reduce metal uptake into plants), and the need to try to take into account degradation mechanisms which may act over very prolonged periods of time.

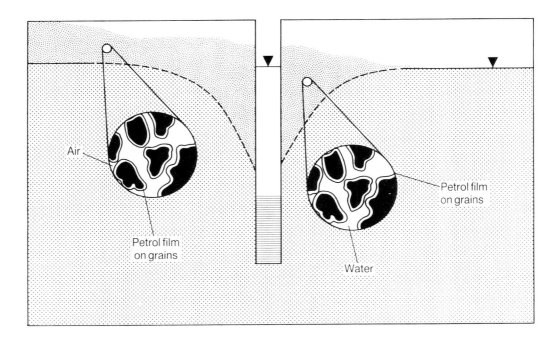

Note: The point of concern is that the cone of depression will contain trapped residual petrol below the water-table which will become a continuous source of contamination that will persist even when the extraction well is turned off. The extent of the contamination that is generated by the residual petrol in the cone of depression may exceed that generated by the petrol resting in place above the saturated zone prior to the onset of pumping.

Figure 7.4 *Zone of residual contamination caused by pumping: pumping creates a zone of depression to trap light organics such as petroleum for removal but also leaves residues below the water table*

7.5.2 In-situ remediation of soils and similar materials

The methods to be employed during remediation and on completion depend on the processes being employed (see Table 7.2).

Proving that a site is clean is much more difficult than establishing that it is contaminated (a single sample may be sufficient in the latter case — see Section 4). Decisions will be required about the sampling and analytical strategies to be applied to:

- process trains (e.g. extracted water containing excess treatment agents and reaction products, extracted gas)
- the ground and groundwater following treatment.

The former should help to determine if the process is working as planned, the stage that it has reached and provide the information for mass balance calculations.

Table 7.2 *Monitoring and compliance testing of some in-situ treatment methods*

Method	Examples of requirements
Soil vapour extraction	
(i) soil in unsaturated zone	Concentration of volatile contaminants remaining in the ground measured at fixed locations
	Oxygen and carbon dioxide concentrations (indicative of rate of air entry and/or microbial activity) in the ground at fixed locations
	Concentrations of volatile contaminants and carbon dioxide (indicative of microbial activity) in extracted gas
	Mass balance
	Determination of soil concentrations on 'completion' of works
(ii) other measures	Monitoring effect of treatment on residual floating layer
	Monitoring groundwater concentrations
	Monitoring for radon (note radon is nearly always present in groundwater and could become concentrated in pollution control media — see Box 7.4)
Microbial treatment	Concentrations of volatiles/semi-volatiles remaining at fixed locations
	Concentrations of possible volatile intermediates, carbon dioxide and oxygen at fixed locations
	Concentrations of contaminants, intermediates, and micro flora and fauna in extracted water
	Determination of soil concentrations on completion of works
Soil washing/flushing	
(i) treatment zone entering the ground	Quality and quantity of flushing water
	Quality of flushing water at fixed intermediate locations
	Quality and quantity of water at discharge/extract point
	Determination of soil concentrations on completion of works
(ii) other measures	Monitoring around site for 'escape' of contaminants and treatment water
	Monitoring effect of treatment on physical properties of soil
Chemical treatment	
(i) treatment zone	Composition and quantity of treatment fluids
	Composition and quantity of discharge/extracted fluid
	Determination of concentrations of contaminants and treatment agents remaining in the soil on completion of works
(ii) other measures	Monitoring around site for 'escape' of contaminants, process residuals or excess treatment agents
	Monitoring for gas generation (e.g. reactions between acids and contaminants or naturally occurring minerals)
	Monitoring effect of treatment on physical properties of soil

For the latter, an investigation strategy must be determined involving decisions on:

- sampling pattern
- nature of samples
- number of samples
- analytical strategy
- statistical basis for determining compliance
- and quality control and quality assurance requirements.

A systematic sampling programme will be required: one of the more rigorous forms is preferable (see Section 4). The statistical basis of sampling, as discussed by Ferguson, is relevant[17,18].

The basis for deciding whether there is compliance with pre-set remedial objectives must be decided before the investigation is carried out. Similar considerations apply to those discussed above in relation to ex-situ process-based methods. Some examples of compliance requirements are described in Boxes 7.4 to 7.6.

Box 7.4 *Evaluation of soil vapour extraction project*[19-21]

The Thomas Solvent Raymond Road (TSRR) site was used for solvent repacking and distribution. Solvents were stored in 21 underground storage tanks found to be leaking. Initial investigations revealed concentrations of volatile organic compounds (VOCs) of 100 mg/kg in groundwater and 1800 mg/kg in soils. It was estimated that the total mass of organics in groundwater and soil was 1800 kg and 770 kg respectively. This was a serious underestimate revised to 5800-7500 kg total following stage (i) of the quality assurance testing plan. After about 18 months of full scale operation over 18 000 kg had been removed, the extra amount in part being attributed to a floating layer not discovered during initial investigations. This underestimate led to unnecessarily high costs (see below). The form of remedial action selected for the site was groundwater extraction in conjunction with soil vapour extraction. It was estimated that groundwater contamination could be reduced to 100 microgram per litre in 3 years and the total contaminant mass reduced by 98% in 1.5 years.

The contractor was set a performance goal of 10 mg/kg 'total VOCs' in the unsaturated zone soils. A three stage sampling plan was devised to demonstrate performance:

(i)   sampling during installation of extraction wells
(ii) sampling at the assumed mid-point of the treatment
(iii) sampling at the point when the contractor was of the opinion that the performance target (10 mg/kg total VOCs) had been met.

Confirmation of compliance required that independent investigation demonstrated :
- no samples taken at any point in the volume of contaminated soil exceeded 10 mg/kg
- no more than 15% exceeded 1 mg/kg.

Notes: *At the time of writing the results of the phase (ii) and (iii) sampling operations are not available.*

Underestimation of the quantity of volatile organics present led to the use of activated carbon to trap volatiles. This was an unnecessarily expensive option: other more economic treatment systems could have been employed for the larger volumes found in practice. In addition, there was a build up of radon in the activated carbon.

Box 7.5 *Monitoring and determination of effectiveness of soil flushing to remove cadmium*[22]

Land (6000 m² in area) contaminated with cadmium adjacent to the site of a photopaper factory was cleaned by in-situ washing with dilute hydrochloric acid. Cadmium was extracted from the extracted wash solution by ion-exchange. The 30 000 m³ soil to be treated was estimated to contain 725 kg of cadmium (an average of about 15 mg/kg assuming a bulk density of 1500 kg/m³). The target concentration set was 2.5 mg/kg with a specified percentage of samples permitted in the range 2.5-5.0 mg/kg. The system was monitored by determining cadmium concentrations in the extracted percolate. Soil concentrations were measured when, on the basis of percolate monitoring data, it was believed that cadmium removal was sufficiently complete. One small area provided samples above the upper 5 mg/kg limit: it was possible to ascribe this to the presence of wastes that were not easily leached. These wastes were subsequently removed. Setting aside the samples from this small area, 91% had concentrations less than 1 mg/kg, 97% below 2.5 mg/kg and 100% below 5 mg/kg (a total of 183 samples were taken (about 1 sample/164 m³). At the time treatment was stopped, about 440 mg/kg had been removed (a residual concentration of 1 mg/kg would amount to about 50 kg cadmium left in the soil). Figures 7.5 and 7.6 show monitoring data for one of the cells.

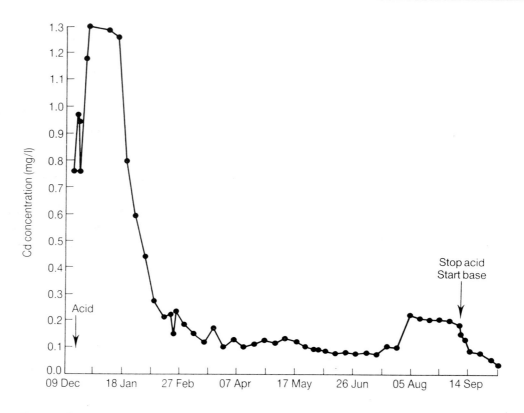

Figure 7.5 *Cadmium concentration in pumped percolate from in-situ soil washing operation (see Box 7.5)*

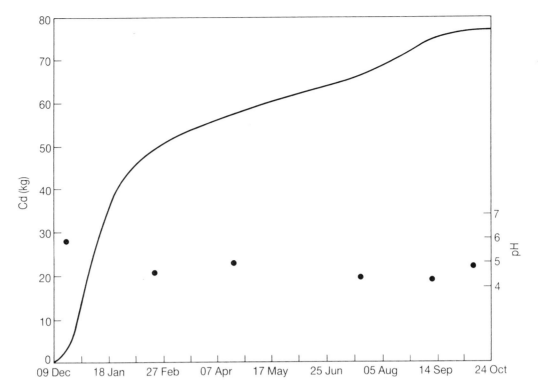

Figure 7.6 *Cumulative quantity of cadmium removed and pH of the percolate from in-situ soil washing operation (see Box 7.5)*

7.5.3 In-situ treatment of groundwater/liquid phases

Groundwater monitoring wells must be installed to determine whether in-situ groundwater treatment has been successful. They should be sited so that the extent of the plume, the degree of contamination and the quality of water entering the treatment zone/site can be regularly monitored, i.e.:

- in part systematically (e.g. on a grid pattern)
- on the basis of the established hydrogeology of the site
- on and off site as appropriate.

Long-term monitoring, perhaps extending well beyond the point when treatment is considered complete, will be required for the reasons explained in Section 7.4, and illustrated in Figures 7.1 and 7.2.

As a measure of performance, the quality of extracted water is unlikely to be sufficient on its own for similar reasons to those outlined above in connection with ex-situ treatment. Again, the (first) measure of success might be the achievement of (say) 'drinking water standards' at specified locations away from the site.

Box 7.6 *Importance of monitoring soil vapour extraction systems for evidence of microbial activity*[23]

Figure 7.7 shows the calculated total removal of petroleum by combined soil venting and induced microbial activity at a site in the Netherlands. The amount volatilised was determined by direct measurements of the extracted gas. The amount removed by microbial action was calculated (after correction using respiration data from uncontaminated reference soils) assuming a zero rate reaction from the rates of oxygen consumption and carbon dioxide production. It can be seen from the Figure that as time passes and the more volatile components are removed, microbial degradation assumes greater importance. After one year the volatilisation and microbial degradation rates were estimated to be 21 mg carbon/kg soil/day and 7 mg carbon/kg/day respectively.

Monitoring volatile gas concentrations in the extracted gas only, which would progressively fall (see Figures 7.8 and 7.9), could lead to premature curtailment of a treatment system before its full potential has been reached. In this case, microbial degradation of toluene was estimated at 2 mg carbon/kg/day.

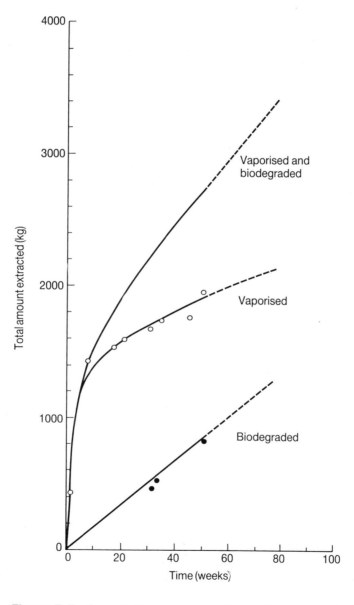

Figure 7.7 *Cumulative amounts of petrol removed during soil vapour extraction: relative contributions of vaporisation and biodegradation (see Box 7.6)*

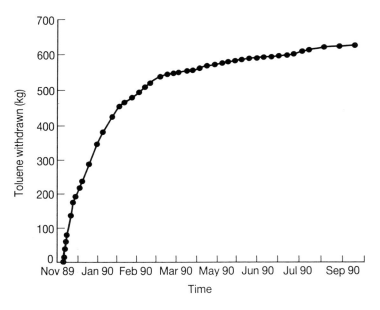

Figure 7.8 *Cumulative amounts of toluene withdrawn in soil vapour extraction system (see Box 7.6)*

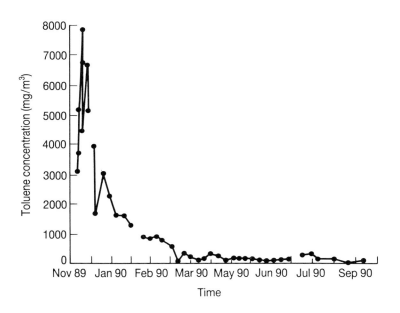

Figure 7.9 *Concentration of toluene in withdrawn soil vapour (see Box 7.6)*

7.6 MONITORING CONTAINMENT METHODS

7.6.1 General requirements

Proper design of containment systems requires that measures are taken at the time of installation to ensure that all design requirements have been met, and that provision is made to monitor long-term performance. Containment systems are engineering works, and, as in all works (e.g. bridges, dams, tunnels) where health and safety or environmental protection are essential, regular monitoring and, where appropriate, maintenance programmes should be put in place. These must address both technical aspects (e.g. methods and frequency of monitoring) and also administrative aspects, e.g.:

- who pays
- who carries out the monitoring
- acceptance/non-acceptance criteria for assessing the results
- what happens to the results
- who has to respond to the monitoring results
- how any essential remedial works to are be paid for.

It is assumed here that appropriate quality assurance programmes will have been applied at the time of installation (see also Volume VI and references 24 and 25) to ensure :

- all materials used meet requisite standards
- the mixes used comply with specified proportions
- materials are placed to the required standards
- materials meet all short-term test requirements
- the basis of design is clear and documented
- expectations regarding long-term behaviour have been properly recorded
- provision has been made for the collection and preservation of any baseline samples and data required to aid judgements about performance in the future.

7.6.2 Cover systems

The performance of cover systems is largely governed by natural processes; they may not be fully tested until many years after installation. At the same time, a number of processes may operate to reduce long-term effectiveness. Most covering systems are multi-functional and no single measure can provide information on the performance of the full range of functions. Each may need to be addressed separately although it may be sufficient to concentrate on a limited number of key functions.

Examples of permanent monitoring measures for containment systems are listed in Table 7.3.

Very few functions, apart from those concerning fluid flow, can be tested at the time of installation. Gas and water flows in collection systems can be monitored for quantity and composition, and checks can be made to confirm that gas is absent from protected areas. Measurements of soil densities and permeabilities can be made at selected locations. Other functions, such as ability to sustain vegetation, prevent erosion, or limit root penetration, cannot be assessed until some years after installation.

Monitoring activities that might be appropriate include:

- regular (normally two or more times a year to allow for natural changes in growth patterns) and systematic site inspections primarily involving visual inspection but also the use of simple instrumentation (e.g. gas detectors)
- site inspections following 'unusual' events (e.g. periods of prolonged rainfall or drought, severe storms)
- sampling of plants and immediate surface (litter) layers

- monitoring of physical properties such as in-situ soil suction and water levels
- monitoring of fluids (gas and water) from fixed points.

Table 7.3 *Possible permanent monitoring measures for containment systems*

Containment system	Examples of measures
General	Variations in water table Extreme climatic events Rainfall Gas concentrations and pressures Ground temperatures Groundwater quality
Covers	Variation of soil moisture content at chosen levels Contaminant concentrations in soil moisture
Vertical barriers	Groundwater quality 'outside' the barrier Resistivity measurements 'outside' the barrier Groundwater levels inside and outside the barrier Potentials across barrier (e.g. electrochemical) Gas concentrations 'inside' and 'outside'
Horizontal in-ground barriers	Groundwater quality around site Groundwater levels within site

Items to be covered by regular inspections are listed in Box 7.7. Because of the long time period over which monitoring is likely to be required, it is essential that adequate base-line data are obtained and reference samples and photographs preserved as necessary.

Box 7.7 *Inspection of cover systems*

Site inspection should embrace :
- the extent, nature and quality of vegetation
- signs of animal activity
- evidence of erosion or settlement
- signs of water logging
- signs of leachate
- signs of cracking of clay covers
- the condition of buildings and services
- gas emissions.

A photographic record (from specified vantage points) and summary of recent weather conditions should also be made.

Physical sampling of cover materials and underlying strata may be difficult. They should not breach the cover system, e.g. by damaging a membrane or clay sealing layer. However, with care, it should be possible to sample the upper soil layers of a cover system without causing permanent damage. Some studies have been made of completed systems to determine the quality of installation and their effectiveness in practice[26].

Cairney[2,24] has described a simple monitoring installation (see Figure 7.10) that can be used to record variations in soil moisture/content at chosen levels, the position of the groundwater table, and soil moisture/contamination concentrations at particular levels to confirm that a cover is behaving according to predictions (this assumes that the cover has been designed using an appropriate modelling system). According to Cairney[24]

> *'the monitoring necessary is neither extensive, requires only equipment that is commercially available, and can easily be undertaken.'*

According to Cairney monitoring of this type need extend for only a few years. Monitoring data confirming model predictions should increase confidence that the cover is behaving (shortly after installation) as expected. It does not remove the need for longer-term surveillance since if extended over a longer-time period, monitoring might give warning of unexpected deviations from predicted behaviour that might not become manifest (e.g. obvious damage to plants) for many years.

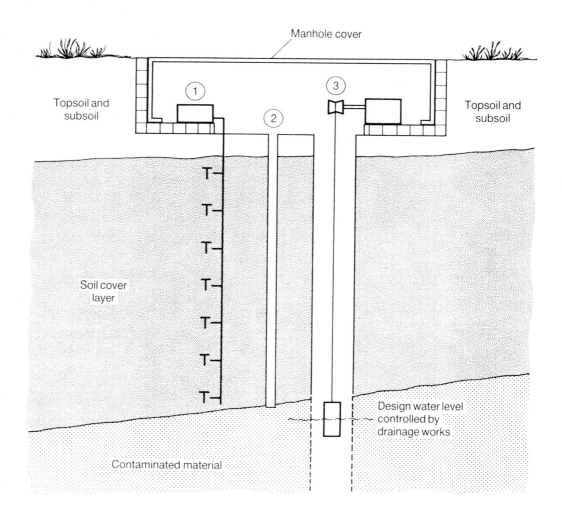

(1) tensiometer array connected to data logger to measure soil suction at points in cover profile
(2) aluminium access tube for neutron probe to obtain soil moisture values
(3) water level, water sampling and gas monitoring borehole (as appropriate) – can also be linked to data logger.

Figure 7.10 *Monitoring system for cover systems (after Cairney[2]).*

The fact that changes are likely to be gradual rather than sudden is an advantage. However, it does mean that apparently unproductive monitoring will have to be continued over many years, and it may be very difficult to judge when deterioration has occurred to such an extent that remedial work is required. For example, if plants die the need for action may be clear (once it is established that contamination, rather than say shortage of water or misuse of herbicides is the cause). In contrast, if average concentrations of cadmium in selected food crops rise by a small amount the judgement will be more difficult. In the latter case, it may be necessary to sample food crops over a number of years to be sure that the effect is real and even then a judgement must be made as to whether this poses any risk to health.

7.6.3 Vertical and horizontal in-ground barriers

When a vertical barrier is intended to prevent the bulk migration of water or gas, standard monitoring wells sited on the outside of the barrier can be used to determine effectiveness at the time of installation. There should be a marked reduction in flows or a decrease in the concentrations of contaminants of concern. Monitoring over a significant period of time may be required to be sure that there are no flaws in the system as installed. Contaminated groundwater may only disperse slowly away from the newly installed wall and gas trapped in the ground may take time to disperse. Indeed, as the barrier may be deliberately designed to create net flow of groundwater inwards, contaminants may persist outside the barrier for a very long time. Great care is required in the design of any monitoring system and considerable time may pass before trends become apparent. As with other measures aimed at improving groundwater quality, success (at least initially) may be judged by whether concentrations in the groundwater at some remote location now achieve acceptable levels.

The major component of a long-term monitoring system is a series of monitoring wells installed on the 'clean' side of the barrier. Data on water levels will also be required from the 'dirty' side, e.g. to ensure that unacceptable rises are not occurring. Other possibilities include the installation of permanent resistivity measuring equipment (perhaps for example based on the British Geological Survey Rescan system) in natural ground or in the barrier itself when this is constructed of cement-bentonite or similar soft materials. The construction of tests cells at the time of installation (see Figure 7.11) may also be helpful. These can be used to measure in-situ permeabilities etc., including the integrity of joints. Guidance being developed in relation to long-term monitoring of operating and closed landfill sites[27-29] may be relevant. However, there is still relatively little experience in the United Kingdom on the installation and operation of long-term monitoring systems.

If the barrier is also intended to control gas migration, comparable data from 'inside' and 'outside' will be required.

Sampling of construction materials once placed should only be done if absolutely essential. Any such exercise will pose a risk of damage to the installed barrier. There may, however, be scope for the installation of sacrificial lengths of barrier under similar field conditions, and parallel long-term laboratory tests.

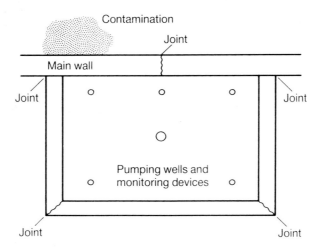

Figure 7.11 *Testing cement-bentonite barrier wall*

7.7 EVALUATION OF REMEDIAL METHODS

7.7.1 General considerations

The evaluation of remedial methods involves gathering far more data than simple compliance or performance testing. The aim typically is to make some general statements about the effectiveness, applicability, and limitations of the method or other remedial measure. It involves establishing how the method works, and also assessing potential health and safety and environmental impacts.

Evaluation studies are expensive and should not be undertaken without a great deal of planning and preparation. Inadequately designed studies may give misleading results. The involvement of organisations with specialist skills and independent of interested parties is recommended.

Guidance produced by EPA is relevant. Guidelines for the evaluation of processes have also been produced in the Netherlands.[3]

7.7.2 Removal and destruction methods

The evaluation of methods which remove or destroy contaminants usually requires a consideration of factors governing both the operation of the process, and its practical application on site. These include:

- responses to changes in feedstock (e.g. contaminants, contaminant concentrations, water content, particle sizing)
- waste emissions (gases, solids, liquids)
- formation of possibly (more) hazardous intermediates or final products
- identification of materials handling problems and requirements
- variability of product under different operating conditions
- suitability of product (e.g. treated soil) for re-use
- health and safety implications
- information needs

- site and operational requirements
- environmental impacts
- regulatory aspects.

7.7.3 Other methods

Methods which do not result in the removal or destruction of contaminants include:

- amendment of surface layers
- containment
- stabilisation/solidification
- in-situ vitrification.

The most important aspects of any evaluation process for these methods are:

- to properly characterise the material(s) to be contained or treated, and
- to accept that they essentially constitute long-term processes and therefore involve long-term research (possibly lasting decades) to fully establish performance, applicability and limitations.

The latter can be addressed to some extent by shorter-term accelerated laboratory tests but test periods of several years may still be required. Long-term studies therefore need careful design and some security of funding if they are to provide meaningful results.

7.7.4 Demonstration projects

Demonstration projects are evaluation studies carried out on a full (occasionally pilot) scale to demonstrate that a particular technology, or (rarely) some other remedial measure, will work under defined practical conditions. There will usually be little doubt about the technical success of the main treatment process (i.e. that it will achieve compliance with agreed remedial objectives) but there may be little information available on emissions for example, or on materials handling aspects.

Demonstration projects have been shown[12] to be of great value in establishing the merits of particular technologies and providing independent evidence of their capabilities. They are the basic mechanism employed by the US Environmental Protection Agency to encourage the use of innovative process-based technologies within the Superfund program and more generally in site remediation.

Such projects need to be carefully designed and planned. A poor demonstration project can have a very negative impact on the acceptability of a technology. Any monitoring or evaluation activities should be carried out by specialist organisations independent of the interested parties.

REFERENCES

1. STIEF, K. The long-term effectiveness of remedial measures. In: *Contaminated Land: Reclamation and Treatment*. Smith, M.A. (ed.) Plenum (London), 1985, pp 13-36
2. CAIRNEY, T. Long-term monitoring of reclaimed sites. In: *Reclaiming Contaminated Land*. Cairney, T. (ed.) Blackie (Glasgow), 1987, pp 170-180

3. INTRON MATERIAL TESTING & CONSULTING. *Opleveringscontrole van Procesmatg Gereinigde Grond en Groundwater, (Testing of treated soil and groundwater)* Maastricht, 1986 (in Dutch)

4. SMITH, M.A. International study of technologies for cleaning-up contaminated land and groundwater. In: *Proceedings of a Conference LAND REC '88*. Durham County Council (Durham), 1988, pp 259-266

5. UNITED STATES ENVIRONMENTAL PROTECTION AGENCY. *Site Program Demonstration Test: Shirco Pilot-scale Infrared Incineration System at the Rose Township Demode Road Superfund Site.* Technology Evaluation Report. USEPA Risk Reduction Engineering Laboratory (Cincinnati), 1989, EPA/540/5 89/007a

6. UNITED STATES ENVIRONMENTAL PROTECTION AGENCY. *CF Systems Organics Extraction System, New Bedford, Massachusetts.* Technology Evaluation Report. USEPA Risk Reduction Engineering Laboratory (Cincinnati), 1990, EPA/540/5 90/002

7. UNITED STATES ENVIRONMENTAL PROTECTION AGENCY. *Site Program Demonstration of the Ultrox International Ultraviolet Radiation/Oxidation Technology.* Technology Evaluation Report. USEPA Risk Reduction Engineering Laboratory (Cincinnati), 1989, EPA/540/5 89/012

8. UNITED STATES ENVIRONMENTAL PROTECTION AGENCY. *Chemfix Technologies Inc. Solidification/Stabilization Process, Clackamas, Oregon.* Technology Evaluation Report, Volume I. USEPA Risk Reduction Engineering Laboratory (Cincinnati), 1989, EPA/540/5 89/011a

9. UNITED STATES ENVIRONMENTAL PROTECTION AGENCY. *Site Program Demonstration Test – The American Combustion Pyreton Thermal Destruction System at the US EPA's Combustion Research Facility.* Technology Evaluation Report. USEPA Risk Reduction Engineering Laboratory (Cincinnati), 1989, EPA/540/5 89/008

10. UNITED STATES ENVIRONMENTAL PROTECTION AGENCY. *Site Program Demonstration Test: Terra Vac In Situ Vacuum Extraction System, Groveland, Massachusetts.* Technology Evaluation Report, Volume I. USEPA Risk Reduction Engineering Laboratory (Cincinnati), 1989, EPA/540/5 89/003a

11. SAMSON, R. GREER, C.W. and HAWARI, J. *Demonstration of a new biotreatability protocol to monitor a bioprocess fro the treatment of contaminated land.* National research Council of Canada, Biotechnology Research Institute (Ottawa), 1992

12. *Final Report NATO Committee on the Challenges of Modern Society Pilot Study: Demonstration of Remedial Action Technologies for Contaminated Land and Groundwater, Volume 1*. USEPA, Risk Reduction Engineering Laboratory (Cincinnati), 1993, EPA/600/R 93/012a

13. YLAND, M. W. F. and SOCZO, E. R. Practical evaluation of a soil treatment plant. *Final Report NATO Committee on the Challenges of Modern Society Pilot Study: Demonstration of Remedial Action Technologies for Contaminated Land and Groundwater, Volume 2 – Part 2*. USEPA, Risk Reduction Engineering Laboratory (Cincinnati), 1993, EPA/600/R-93/012c, pp 911 -928

14. STEGEMANN, J.A. and COTE, P.L. *Investigation of Test Methods for Solidified Waste Evaluation – A Cooperative Program*. Report EPS 3/HA/8. Wastewater Technology Centre (Burlington, Ontario), 1991

15. HINSENVELD, M. A sound and practical method to determine the quality of stabilization. In: *Proceedings of the First International NATO/CCMS Conference on the Evaluation of Demonstrated and Emerging Technologies for the Treatment and Cleanup of Contaminated Land and Groundwater (Phase II), (Budapest)*. USEPA Risk Reduction Engineering Laboratory (Cincinnati), 1992

16. HINSENVELD, M. *Leaching and volatilization from cement stabilized wastes*. Kluwer (Dordrecht), 1993

17. FERGUSON, C. A statistical basis for spatial sampling of contaminated land. *Ground Engineering*, 1992, (June), 34-38
18. FERGUSON, C. *Sampling Strategies for Contaminated Land.* CLR Report No. 4. Department of the Environment, (London), 1994
19. TANAKA, J.C. Soil vapour extraction system: Verona Well Field Superfund Site, Battle Creek, Michigan USA. In: *Proceedings of the First International NATO/CCMS Conference on Demonstration of Remedial Action Technologies for Contaminated Land and Groundwater (Washington DC).* USEPA (Cincinnati), 1987, pp 182-187
20. ANON. Soil treatment by vacuum extraction at Verona Well Field. In: *Proceedings of the Second International NATO/CCMS Conference on Demonstration of Remedial Action Technologies for Contaminated Land and Groundwater (Bilthoven).* USEPA (Cincinnati), 1988, pp 217-240
21. GUERRIERIO, M.M. In-situ vacuum extraction Verona Well Field Superfund Site, Battle Creek, Michigan. *Final Report NATO Committee on the Challenges of Modern Society Pilot Study: Demonstration of Remedial Action Technologies for Contaminated Land and Groundwater, Volume 1.* USEPA, Risk Reduction Engineering Laboratory (Cincinnati), 1993,. EPA/600/R-93/012a, pp 1032 -1052
22. URLINGS, L.G.C.M. et al. In situ cadmium removal. *Final Report NATO Committee on the Challenges of Modern Society Pilot Study: Demonstration of Remedial Action Technologies for Contaminated Land and Groundwater, Volume 1.* USEPA, Risk Reduction Engineering Laboratory (Cincinnati), 1993,. EPA/600/R-93/012a, pp 1135 – 1156
23. URLINGS, L.G.C.M., COFFA, S. and van VREE, H.B.R.T. In situ vapour extraction. In: *Proceedings of the Fourth International NATO/CCMS Conference on Demonstration of Remedial Action Technologies for Contaminated Land and Groundwater (Angers).* USEPA (Cincinnati), 1990
24. CAIRNEY, T. and SHARROCK, T. Clean cover technology. In: *Contaminated Land: Problems and Solutions.* Cairney, T. (ed.) Blackie (London), 1993, pp 84-110
25. JEFFERIS, S.A. In-ground barriers. In: *Contaminated Land: Problems and Solutions.* Cairney, T. (ed.) Blackie (London), 1993, pp 111-140
26. RICHARDS MOOREHEAD & LAING. *An Assessment of the Effectiveness of the Methods and Systems used to Reclaim Contaminated Sites in Wales.* Report for the Welsh Office, 1988
27. DEPARTMENT OF THE ENVIRONMENT. *Landfilling Wastes.* Waste Management Paper No 26. HMSO (London), 1986
28. DEPARTMENT OF THE ENVIRONMENT. *Landfill Gas.* Waste Management Paper No. 27 (2nd Edition). HMSO (London), 1991
29. DEPARTMENT OF THE ENVIRONMENT/WELSH OFFICE/SCOTTISH OFFICE. *Landfill completion.* Waste Management Paper No. 26A, HMSO (London), 1993

Further reading

SCG. *Inspection of remediated ground and ground to be remediated on behalf of the SCG (Onderzoek te reinigen grond en gereinigde grond t.b.v. het SCG).* SCG (Netherlands), 1992

Appendix 1 Guidance documents on site investigation

A1.1 KEY DOCUMENTS

A1.1.1 BS DD175:1988 Draft for Development Code of Practice for the identification of contaminated land and its investigation

BS DD 175 provides information and guidance on topics including:

- definitions, objectives and strategies
- planning issues and site investigation procedures, including sampling and analysis programmes and techniques
- the assessment of findings
- reporting
- health and safety issues.

The life of the Draft for Development was extended in 1990. The British Standards Institution (BSI) is currently (January 1995) considering a strategy for the revision of DD 175 on completion of work by the International Organisation for Standardisation (ISO) to provide guidance on site investigation and sampling.

A1.1.2 BS 5930:1981 Code of Practice for Site Investigations

The British Standard *Code of Practice for Site Investigations* deals with the investigation of sites for the purposes of assessing their suitability for the construction of civil engineering and building works, and as a means of obtaining information on other characteristics of the site which may affect the design or construction of the works, or the security of neighbouring land and property. It provides detailed information on:

- technical, legal and environmental factors affecting the suitability of the site and the design of the works
- planning the investigation
- methods of ground investigation including excavation, boring, sampling, probing and testing in boreholes; field testing; laboratory analysis
- report preparation
- terminology and classification systems.

BS 5930 is currently (January 1995) being revised. The current draft includes an Appendix dealing with contamination in the context of construction-related investigations.

A1.1.3 CIRIA SP25 Site investigation manual

The CIRIA *Site investigation manual* complements BS 5930 with guidance on the coordination of site investigation, supervision of site works and aspects of foundation design. As with BS 5930, the Manual treats investigation as part of the evaluation of a potential construction site and its suitability for the proposed works. Detailed guidance is given on:

- methods and equipment for exploration and sampling of soils and weak rocks
- in-situ testing, including geophysical exploration
- laboratory testing of soils and weak rock
- groundwater and permeability
- soil and rock description, classification and logging.

A1.2 OTHER GUIDANCE DOCUMENTS

A1.2.1 American Petroleum Institute (API)

API Publication 1628
A guide to the assessment and remediation of underground petroleum releases
API (Washington DC), 1989

A1.2.2 American Society for the Testing of Materials (ASTM)

Standard D 5092
Standard practice for the design and installation of groundwater monitoring wells in aquifers
ASTM (Philadelphia, Pennsylvania), 1990

Standard D 5088
Standard practice for the decontamination of field equipment used at non-radioactive waste sites
ASTM (Philadelphia, Pennsylvania), 1990

A1.2.3 Association of Geotechnical Specialists (AGS)

Collateral warranties
AGS (Camberley) 1993

Quality management in geotechnical engineering - a practical approach
AGS (Camberley) 1990

Safety manual for investigation of sites
AGS (Camberley)

Electronic transfer of geotechnical data from ground investigation
AGS (Camberley) being revised 1994

A1.2.4 British Drilling Association (BDA)

Code of safe drilling practice Part 1: surface drilling
BDA (Brentwood), 1981

Guidelines on the drilling of landfill, contaminated land and adjacent areas
BDA (Brentwood), 1991

A1.2.5 British Standards Institution

BS DD 175:1988
Code of practice for the identification of potentially contaminated land and for its investigation

BS 5930:1981
Code of practice for site investigations

BS 1377:1990
Methods of test for soils for civil engineering purposes

BS 6068: *Water Quality: Part 2: Chemical and biochemical methods*

BS 6068: *Water Quality: Part 4: Microbiological methods*

BS 6068: *Water Quality: Part 5: Biological methods*

BS 6068:Part 6: Section 6.1:1981 (confirmed 1990) (ISO 5667/1-1980)
Water Quality: Part 6.1 Sampling: Section 6.1 Guidance on the design of sampling programmes

BS 6068:Part 6: Section 6.2:1991 (ISO 5667/2-1991),
Water Quality: Part 6 Sampling: Section 6.2 Guidance on sampling techniques

BS 6068:Part 6: Section 6.3:1986 (ISO 5667/3-1985)
Water Quality: Part 6 Sampling: Section 6.3 Guidance on the preservation and handling of samples

BS 6068:Part 6: Section 6.4:1987 (ISO 5667-4:1987)
Water Quality: Part 6 Sampling: Section 6.4 Guidance on sampling from lakes, natural and man made

BS 6068:Part 6: Section 6.6:1991 (ISO 5667-6:1991)
Water Quality: Part 6 Sampling: Section 6.3 Guidance on sampling of rivers and streams

BS 6068: Part 6: Section 6.11: 1993 (ISO 5667-11)
Water Quality: Part 6 Sampling: Section 6.11 Guidance on the sampling of groundwaters

BS 6068:Part 6: Section 6.12:1993 (ISO CD 5667-12)
Water Quality: Part 6 Sampling: Section 6.12 Guidance on sampling of sediments

BS 7755
Soil Quality: Part 1 Vocabulary
    Part 2 Sampling
    Part 3 Chemical methods
    Part 4 Biological methods
    Part 5 Physical methods

(Note: ISO soil quality standards will be published in the BS 7755 series. For listing of pending and published ISO standards see A1.2.20 below and Appendix 7.)

A1.2.6 Building Research Establishment (BRE)

Digest 275
Fill: Part 2: Site investigation, ground improvement and foundation design

Digest 318
Site investigation for low-rise building: desk studies

Digest 322
Site investigation for low-rise building: procurement

Digest 348
Site investigation for low-rise building: the walk over survey

Digest 363
Sulphate and acid resistance of concrete in the ground

Digest 381
Site investigation for low-rise building: trial pits

Digest 383
Site investigation for low-rise building: soil description

IP2/87
Fire and explosion hazards associated with the redevelopment of contaminated land

IP3/89
Subterranean fires in the UK — the problem

D Crowhurst
Measurement of gas emissions from contaminated land

BRE Publications are available from the Bookshop, Building Research Establishment, Garston, Watford WD2 7JR

A1.2.7 Canadian Council of Environment Ministers (CCME)

National guidelines for decommissioning industrial sites
CCME (Winnipeg) 1991
Reprt CCME-TS/WM=TRE 013E

Guidance manual on sampling, analysis and data management for contaminated sites, Volume 1: Main report
CCME (Winnipeg) 1993
Report CCME EPC-NCS 62E

Guidance manual on sampling, analysis and data management for contaminated sites, Volume 2: analytical method summaries
CCME (Winnipeg) 1993
Report CCME EPC-NCS 66E

Subsurface assessment handbook for contaminated sites
CCME (Winnipeg) 1994
Report CCMe EPC-NCSRP-48E

A review of whole organism bioassays for assessing the quality of soil, freshwater sediment and freshwater in Canada
(Report to CCME 1992)

A1.2.8 Chemical Industries Association

Contaminated land and land remediation - guidance on the issues and techniques
CIA (London) 1993

A1.2.9 Confederation of British Industry

Guidelines for business to deal with contaminated land
CBI (London) 1994

A1.2.10 Construction Industry Research and Information Association (CIRIA)

SP 25, 1989
A.J.Weltman & J M Head
Site investigation manual

SP 45, 1986
Recommendations for the procurement of ground investigation

SP 73, 1991
Role and responsibility in site investigation

SP 78, 1991
Building on derelict land

SP79, 1992
Methane and Associated Hazards to Construction: a bibliography

Report 130, 1993
Methane: Its occurrence and hazards in construction

Report 131, 1993
The measurement of methane and other gases from the ground

FR/CP/14, 1993
Methane investigation strategies

FR/CP/22
Interpreting measurements of gas in the ground

FR/CP/23
Risk assessment for methane and other gases in the ground

FR/CP/25
Remedial treatment of contaminated land using in-ground barriers, liners and cover systems

Report 132
A guide to safe working practices for contaminated sites

Report 113, 1986
Control of groundwater for temporary works

A1.2.11 Department of the Environment

Waste Management Paper 26A
Landfill completion
HMSO (London) 1993

Waste Management Paper No 27
Landfill gas
HMSO (London) 1991

Mineral Planning Guidance 12
Treatment of disused mine openings and availability of information on mined ground
HMSO (London) 1994

CLR 1
A framework for assessing the impact of contaminated land on groundwater and surface water (2 vols)
DoE (London) 1994

CLR 2
Guidance on the preliminary inspection of contaminated land (2 vols)
DoE (London) 1994

CLR 3
Documentary research on contaminated land
DoE (London) 1994

CLR 4
Sampling strategies for contaminated land
DoE (London) 1994

CLR 5
Information systems for contaminated land
DoE (London) 1994

A1.2.12 Department of Transport

Specification and method of measurement for ground investigation
HMSO (London) 1987

A1.2.13 Friends of the Earth

Buyer beware: a guide to finding out about contaminated land
FOE (London), 1993

A1.2.14 Health and Safety Executive

HS(G) 36
Disposal of explosives waste and the decontamination of explosives plant
HMSO (London) 1987

HS(G) 66
Protection of workers and the general public during development of contaminated land
HMSO (London) 1991

HS(G) 46
Avoiding danger from underground services
HMSO (London) 1989

A1.2.15 Highways Agency

Advice note: Site investigation on contaminated land for highway works
HA (London) - draft 1994

A1.2.16 Institute of Petroleum (IP)

Code of Practice for the investigation and mitigation of possible petroleum-based land contamination
IP/Wiley (London) 1993

A1.2.17 Institution of Civil Engineers

Conditions of contract for ground investigation
Thomas Telford (London) 1983

Design guide on contaminated land
Thomas Telford (London) 1994

A1.2.18 Institution of Environmental Health Officers

Contaminated land: development of contaminated land — professional guidance
IEHO (London)

A1.2.19 Interdepartmental Committee on the Redevelopment of Contaminated Land (ICRCL)

ICRCL 17/79
Notes on the development and after-use of landfill sites

ICRCL 18/79
Notes on the redevelopment of gasworks sites

ICRCL 23/79
Notes on the redevelopment of sewage works and farms

ICRCL 42/80
Notes on the redevelopment of scrapyards and similar sites

ICRCL 59/83
Guidance on the assessment and redevelopment of contaminated land

ICRCL 61/84
Notes on the fire hazards of contaminated land

ICRCL 64/85*
Asbestos on contaminated sites

ICRCL 70/90
Notes on the restoration and aftercare of metalliferous mining sites for pasture and grazing

ICRCL Notes are available from Department of Environment, Publications Dept., Government Building, Block 3 Spur 2, Rm 1, Lime Grove, Eastcote, HA4 8SE.

A1.2.20 International Organisation for Standardisation (ISO)

Standards on soil quality covering terminology etc., sampling, chemical analysis, physical characterisation and biological assessment, are being produced by ISO Technical Committee TC 190. Those related to terminology etc. and sampling are listed below. Those on chemical analysis, physical characterisation and biological assessment are listed in Appendix 7. In due course the ISO standards will be adopted as British Standards in the BS 7755 series. ISO standards are available from the British Standards Institution.

ISO = full ISO standard − usually adopted by BSI as British Standard
DIS = draft international standard
CD = committee draft
WD = working draft
WI = work item

CD 11259
Description of soils and sites

DIS 11074-1
Soil quality − Vocabulary − Part 1: Terms and definitions relating to the protection and pollution of soil

CD 11074-2
Soil quality − Vocabulary − Part 2: Terms and definitions related to sampling

WI 11074-3
Soil quality − Vocabulary − Part 3: Terms and definitions related to risk assessment

WI 11259-2
Soil quality − Codification of soil agronomic and analytical data

CD 10381-1
Soil quality − Sampling − Part 1: Guidance on the design of sampling programmes

CD 10381-2
Soil quality − Sampling − Part 2: Guidance on sampling techniques

CD 10381-3
Soil quality − Sampling − Part 3: Guidance on safety

CD 10381-4
Soil quality − Sampling − Part 4: Guidance on the procedure for the investigation of natural, near-natural and cultivated soils

CD 10381-5
Soil quality – Sampling – Part 5: Guidance on the procedure for the investigation of soil contamination of urban and industrial sites

ISO 10381-6
Soil quality – Sampling – Guidance on the collection, handling and storage of soil for assessment of aerobic microbial processes in the laboratory
(BS 7755: Section 2.6: 1994)

A1.2.21 Loss Prevention Council (LPC)

Pollutant industries
LPC (London) 1992

A1.2.22 Ministry of Agriculture, Fisheries and Food

Sampling soil for analysis
Leaflet 655
MAFF (London), 1983

A1.2.23 Ministry of Housing, Physical Planning and Environment (Netherlands)

Ministerie van Volkshuisvestig, Ruimtelijke Oredening en Mileubeheer
Protocol voor het: Orienterend onderzoek
Sdu Uitgeverij Koninginnegracht (Den Haag) 1993

Ministerie van Volkshuisvestig, Ruimtelijke Oredening en Mileubeheer
Protocol voor het: Nader onderzoek deel 1
Sdu Uitgeverij Koninginnegracht (Den Haag) 1993

A1.2.24 Nederlands Normalisatie-Instituut (NNI)

NVN 5740
Soil: Investigation strategy for exploratory survey (Bodem: Onderzoeksstrategie bij verkennend onderzoek)
NNI (Delft) 1991

A1.2.25 New Jersey Department of Environmental Protection (NJDEP)

Field sampling procedures manual
NJDEP (Trenton NJ) 1988

Field procedures manual for water data acquisition
NJDEP (Trenton NJ) 1980

A1.2.26 Ontario Ministry of the Environment and Energy

Guidance on sampling and analytic methods for site clean-ups in Ontario (Draft) OMEE (Toronto) 1994

A1.2.27 Scottish Enterprise

Requirements for contaminated land site investigations
Scottish Enterprise (Glasgow) 1993

A1.2.28 Site Investigation Steering Group

Site investigation in construction:
  1- Without site investigation ground is a hazard
  2- Planning, procurement and quality management
  3- Specification for ground investigation
  4- guidelines for the safe drilling of landfills and contaminated ground
Thomas Telford (London) 1993

A1.2.29 US Environmental Protection Agency

EPA/600/4-89/034
Handbook of suggested practices for the design and installation of ground-water monitoring wells

EPA/625/6-90/016b
Handbook: Groundwater, Volume II: Methodology

OSWER Directive 9380.0-3.
Guidance document for cleanup of surface tank and drum sites

EPA/600/2-90/061.
State-of-the-art procedures and equipment for internal inspection of underground storage tanks

EPA/540/G-80/004
Guidance for conducting remedial investigations and feasibility studies under CERCLA

US Environmental Protection Agency publications can be obtained from the National Technical Information Service, Springfield VA 22161. Some of the more recent publications are available from the Center for Environmental Research and Information, Cincinnati OH 45268

A1.2.30 Welsh Development Agency

Manual on remediation of contaminated land
WDA (Cardiff) 1993

A1.3 OTHER USEFUL TEXTS

BRIDGES, E.M.
Surveying derelict land
Clarendon Press (Oxford) 1987

CLAYTON, C.R.I, SIMONS, N.E. and MATTHEWS, M.C.
Site investigation — a handbook for engineers
Granada Publishing, 1982

COTTINGTON, J. and AKENHEAD, R.
Site investigation and the law
Thomas Telford (London), 1984

HAWKINS, A.B.(ed.)
Site investigation practice: assessing BS 5930
Geological Society (London)

HEAD, K.
Manual of laboratory testing
Volume 1: Soil classification and compaction tests
Volume 2: Permeability and shear strength
Volume 3: Effective stress tests
Pentech Press: v1 – 1980, v2 – 1982, v3 – 1986

HEWITT C.N.
Methods of environmental data analysis
Elsevier Applied Science (London), 1992

HITCHMAN S.P.
A Guide to the field analysis of groundwater
Fluid Processes Unit, Institute of Geological Sciences 1983

KEITH L.H.
Environmental sampling and analysis: A practical guide
Lewis Publishers (Chelsea MI), 1991

NAYLOR, J.A., ROWLAND, C.D., YOUNG, C.P. and BARBER, C.
The investigation of landfill sites
Water Research Centre Technical Report TR 91. 1978

NIELSEN D.M. (edit)
Practical handbook of ground-water monitoring
Lewis Publishers (Chelsea MI), 1991

STUART A.
Borehole sampling techniques in groundwater pollution studies
Fluid Processes Unit, Institute of Geological Sciences, 1983

Appendix 2 Information sources for desk study

Table A2.1 *Information sources likely to be used in the desk study*

Site records	Drawings, production logs, maintenance records, supplier records, environmental audits, compliance records, COSHH assessment records, R&D activities
Company records	Environmental audit reports, archival information, title deeds
Plant personnel	Plant Manager, Safety Manager, Product Manager, Employees
Maps	O.S., town maps, geological maps, hydrogeological maps, thematic maps, groundwater vulnerability maps, tithe maps
Photographic material	Ground level, aerial (e.g. military photo-reconnaissance)
Local literature	Library local history departments, local newspapers, local specialist societies, clubs, etc.
Directories	Trade, street etc.
The regulatory authorities	Local councils (e.g. environmental health, planning, waste regulation authorities, Hazardous Substances Authorities), National Rivers Authority, Health & Safety Executive, Her Majesty's Inspectorate of Pollution, River Purification Boards, Northern Ireland Department of the Environment
Local knowledge	Residents, former employees etc.
Technical literature	Research and review papers, specialist guidance[1]
The fire and emergency services	Fire certification documents, accident reports
Other organisations	British Coal, Opencast Executive, British Geological Survey, Water, Gas and Power Companies etc.

Note: 1. E.g. Industry Profiles (under preparation by the Building Research Establishment on behalf of DoE)

Table A2.2 *Key information sources for hydrological data*

British Geological Survey	Geological maps and records, hydrogeological maps and well records, reports on applied geology for planning and development, geological memoirs etc.
National Rivers Authority and Water Companies	Well records, water quality data, water and river levels, aquifer properties, flow rates, discharges abstraction and recharges, water vulnerability maps
Local Authorities	Miscellaneous records
Meteorological Office	Daily, monthly and annual rainfall and evaporation records
HMSO	Admiralty charts and tide tables etc., government publications
Site Records	Site investigation drawings, logs, records etc.
Various authorities	Public registers established under the Environmental Protection Act 1990 and other relevant Acts

Table A2.3 *Sources of information on the relationship between site use and nature of contamination*

DOE Building Research Establishment
Industry profiles (to be published)

Interdepartmental Committee on the Redevelopment of Contaminated Land
Guidance notes
(see Appendix 1)

Department of the Environment
Waste Management Papers

Contaminated land reports

CIRIA Report 132
A guide to safe working practices for contaminated sites

Loss Prevention Council
Pollutant industries
LPC (London) 1992

Barry, D.L.
Former iron and steelmaking plants
In: Smith, M.A. (ed.), *Contaminated Land: Reclamation and Treatment* (Plenum, London 1985)

Bridges, E.M.
Surveying derelict land
Clarendon Press (Oxford) 1987

Appendix 3 Possible hypotheses on the distribution of contaminants

An approach to the development of hypotheses for contamination is described in NVN 5740 (NNI 1991) *'Investigation Strategy for Exploratory Survey'*. Typical hypotheses which might be applied to a whole site or to parts of a site (i.e. different hypotheses for different parts of the site) include:

H1: that the site (or defined part of it) is uncontaminated

And if contamination is suspected that it is:

H2: heterogeneously distributed

H3: homogeneously distributed, or

H4: heterogeneously distributed without known point sources.

In the case of **H2**, the contamination has a definite point source or sources (e.g. leaking tanks) within the scale of the investigation (site) and major variations occur in both vertical and horizontal planes.

In the case of **H3**, there is no identifiable point source within the scale of the investigation and there are no major variations in the horizontal plane although there may be marked variations in the vertical plane.

In the case of **H4**, evidence for localised contamination exists (e.g. from on-site observations) but the source is not known.

In practice, because of variations in the physical characteristics of strata, and a variety of sources of contamination on a typical old industrial site, they are frequently:

H5: uniformly heterogeneous (i.e. show great variations in both horizontal and vertical planes with little rational pattern being discernable).

Failing any other sound hypothesis, it is this last that should inform the on-site investigation, whether this is the exploratory phase or the main/detailed phase of the investigation.

Failure to confirm an hypothesis during an investigation indicates that the hypothesis was wrong and a new one needs to be formulated. It does not mean that the site is uncontaminated.

The hypothesis about the spatial distribution of contamination should take into account:

- the nature of the source and the manner in which the contamination has entered the soil (diffuse or spot)

- where in the soil or the groundwater the contamination is likely to be found, taking into account the expected migration processes in both vertical and horizontal directions. These will depend on:

  - the nature of the contaminants (e.g., solubility in water, interaction with clay, interaction with organic matter in soil)

  - soil stratification (nature and permeability of soil/fill, thicknesses of strata)

  - soil/fill characteristics such as pH, Redox potential, organic matter content, and cation exchange capacity

  - the depth to groundwater and conclusions drawn about probable direction of flow, and temporal variations of both of these

  - how long the contamination has been in the ground.

Different substances may require different detection procedures because:

- different substances have different physical properties (e.g. viscosity, solubilities, specific densities, volatilities)
- interact differently with soil components
- are affected differently by factors such as pH and redox potential consequently
- may exhibit markedly different distributions.

NVN 5740 (NNI 1991) makes detailed recommendations on investigation strategies for exploratory investigations (including the minimum number of sampling points to be employed, and number of samples to be taken and analysed) to test the first four hypotheses (H1 to H4) listed above. H5 (uniformly heterogeneous) requires a similar approach to that for a site that is believed to be uncontaminated (H1). Sampling strategies are discussed in detail in Section 4.

Appendix 4 Guidance on specialists

A4.1 SOURCES OF INFORMATION ON CONSULTANTS AND CONTRACTORS

Source	Content/Status
British Geotechnical Society, Geotechnical Directory of the United Kingdom[1]	Includes listings of consultants, contractors and individuals who have achieved a certain level of expertise in geotechnical engineering. This may or may not include contamination. The information contained in the directory in respect of contamination expertise does not claim to be authoritative, albeit many of the organisations who regularly undertake such work are in fact listed.
Royal Society of Chemistry/Institution of Civil Engineers/Institution of Biologists listing of organisations specialising in contaminated land investigations[2].	This document is now seriously out of date
ENDs Directory of Environmental Consultants[3]	Provides information on consulting organisations undertaking work in the area of contaminated land. Note that this and other similar directories are based on information submitted by listed organisations. Listing is not intended, and is not claimed by the authors, to offer any recommendation on the suitability of the organisations for any particular area of work.
National Measurements Accreditation Service (NAMAS)[4]	A directory of laboratories offering services in a number of testing applications, including the analysis of soils and waters for contamination assessment purposes. Laboratories listed have been certified by NAMAS to be competent in the field of testing indicated. Accreditation relates to satisfactory performance in the use of a published analytical method.
Association of Environmental Consultancies	Is expected to publish a list of member organisations claiming competence in contaminated land work
British Drilling Association[5]	Operates a certification scheme for ground investigation drillers
Geological Society Directory[6]	Lists organisations offering geological services: i.e. consulting geologists, geochemists, geophysicists, geotechnical engineers and engineering geologists, civil engineers, and mining engineers. Also lists organisations offering specialist services, e.g. drilling, surveying, materials testing, well testing.

REFERENCES

1. THE BRITISH GEOTECHNICAL SOCIETY. *The Geotechnical Directory of the United Kingdom*, Third Edition. BGS (London), 1982

2. ROYAL SOCIETY OF CHEMISTRY/INSTITUTION OF CIVIL ENGINEERS/INSTITUTION OF BIOLOGISTS. (list of contaminated land specialists). RSC (London)

3. ENDS. *Directory of Environmental Consultants Third Edition (1992/93)*. ENDS (London), 1992

4. NATIONAL PHYSICAL LABORATORY. *NAMAS Directory of Accredited Laboratories*. NPL (Teddington), 1993

5. BRITISH DRILLING ASSOCIATION LTD. *Ground Investigation Drillers Accreditation Scheme*. BDA (Brentwood)

6. THE GEOLOGICAL SOCIETY. *The Geologists Directory*. Seventh Edition. The Geological Society (London), 1994

Appendix 5 Key elements in the NAMAS accreditation scheme

Element	Requirement
General requirements	A single level of service Participation in inter-laboratory comparisons, etc., where appropriate or if directed by NAMAS
Organisation and management	Nominated technical director and deputy Nominated quality manager and deputy Authorised signatories for reports
Quality system	Quality manual, kept up to date Controlled distribution of documents
Quality audit and review	Periodic audit of all activities At least annual reviews Quality checks: • quality control scheme • proficiency tests • use of standard reference materials • replicate testing • retesting of retained samples
Staff	Appropriately qualified Documented training records Permit to work Occasionally named individuals (asbestos, calibration)
Equipment	Individual records for each item Maintenance records Use records where appropriate identified with its calibration status
Traceability and calibration	Measurements traceable to National/International standards Accredited outside calibration agencies Certification of calibration
Test methods	Documented controlled distribution, amendments etc. Accredited by individual documented test methods 'Standard' methods preferred
Laboratory accommodation and environment	Appropriate Controlled access Monitoring and control Housekeeping
Handling of test items	Unique identification Condition on receipt Prevention of deterioration before, during, and after testing Disposal

Element	Requirement
Test records	Retained for at least 6 years Full audit trails Workbooks signed and countersigned Reconstruct report from raw data after 6 years
Test reports	Accurate, clear, unambiguous and objective Identify accredited/non-accredited tests Interpretation outside scope of accreditation
Handling of complaints and anomalies	Documented policy Record all complaints Audit Advise other affected clients
Sub-contracting	Register sub-contractors Sub-contract accredited tests only with client's approval Use NAMAS accredited sub-contractors where possible
Outside support services and supplies	Register of approved suppliers Approval system BS 5750/ISO 9000 preferred

Appendix 6 Investigation techniques

This Appendix provides guidance on some of the more important aspects of investigation techniques for contaminated sites. It is not intended to be fully comprehensive: for detailed guidance reference should be made to the various publications cited. Sampling and investigation methods for soils and similar materials are dealt with in Section 4.

A6.1 GROUNDWATER QUALITY MONITORING

A6.1.1 Drilling methods

Two principal methods are used in the UK :

1. **Light cable percussive boring** ('shell and auger') – used in soils and fill materials, with limited application for weathered or weak rocks.

2. **Rotary drilling** (either open-hole or including core extraction) – used for penetrating rock or other hard strata.

The use of **hollow-stem augers** is increasing the UK. Other drilling methods are listed in Box A6.1. The choice of method depends on the purpose of the borehole, geological conditions and time available. Core extraction is significantly more expensive than open-hole drilling and the need for core samples for testing purposes should be considered at the planning and design stage. If one of the objectives is to provide information on the hydrogeological conditions, including geological structure, secondary permeability etc., core samples should be taken. In contamination investigations core extraction provides a much better quality record of the in-situ conditions for visual inspection and sampling than do other sampling methods.

Box A6.1 *Possible drilling methods for installation of monitoring wells*

Hand augers	Direct mud rotary
Driven wells	Air rotary
Jet percussion	Air rotary with casing driver
Solid-flight augers	Dual-wall reverse-circulation
Hollow-stem augers	Cable tool drilling (light percussion)

Guidance on the selection of drilling methods is given in references 1 and 2. Reference 2 provides a series of matrices relating drilling techniques to a range of design and performance criteria.

In contaminated land applications, particular care should be taken in the selection of drilling methods because of the potential for contamination of clean ground and groundwater. Depending on the technique used, the following precautions should be taken :

- use of appropriate casing, kept coincident with the base of the borehole, at all stages of boring (note casing can be used as a temporary sealing tool when passing through the base of contaminated ground into underlying uncontaminated media)

- use of permanent sealing (bentonite or similar) to prevent the creation of preferential paths for the movement of contaminants and leachates

- measures to prevent extraneous matter falling into the borehole (e.g. use of casing during drilling and sealable/lockable covers on completion)

- care in the use of assisted drilling techniques (see Box A6.2).

Mud-flush and water assisted drilling should be avoided where possible. Other potential limitations of assisted drilling techniques for contaminated land applications are listed in Box A6.2.

Box A6.2 *Examples of limitations of different drilling methods*

- the need to filter the air used in air-flush drilling when monitoring for organic contaminants to avoid contamination of samples with oil from the compressor

- the potential for health and environmental impacts (via the discharge of contaminated water and cuttings from the borehole) during air-flush drilling in highly contaminated ground

- the potential for displacement of volatile compounds (leading to unrepresentative ground conditions) when using air-flush drilling techniques

- the potential for introducing organic substances into the borehole when using foam-flush drilling techniques

---

Useful detailed guidance on the formation of boreholes in the investigation of landfill sites is given in a Water Research Centre Technical Report[3]. The advice provided is generally applicable to the investigation of contaminated land.

Appropriate methods for the collection, treatment or disposal of contaminated liquid and solid wastes arising from drilling operations should be addressed.

A6.1.2 Construction materials and methods

The types of materials and methods used to construct groundwater quality monitoring installations should be selected on the basis of :

- the specific objectives of the investigation taking into account the types of contaminants to be monitored, anticipated concentrations, behaviour in the field etc.

- the duration of monitoring and the potential for chemical attack on in-ground equipment.

It is essential that equipment or materials introduced into or used for backfilling or sealing the installation should not affect the groundwater chemistry. Contamination via cement grouts (high pH) and filter packs that are not chemically inert are common problems. The choice of casing material has been extensively researched, particularly in respect of absorption and desorption effects. Casing and screen materials should be:

- new and certified clean

- selected for resistance to chemical attack

- of sufficient physical strength to withstand installation and development stress

- have minimal effect on the groundwater sample taken.

Resistance to chemical attack is important at high contaminant concentrations
(10^2 to 10^3 × mg/litre); absorption and desorption at lower concentrations (μg/litre).

Use of more than one type of construction material may offer cost benefits. For example, stainless steel or fluoropolymer materials can be used in the sampling location, with PVC materials in non-critical portions of the well. In composite well construction, the use of dissimilar metallic materials should be avoided unless they are electrically isolated.

Specific aspects to be addressed include :

- construction materials
- casings and screens
- annular seals
- surface finish
- well development.

A6.1.2.1 Construction materials

These should not interact with or otherwise modify (e.g. through absorption, leaching or other processes) the contamination present in the ground and groundwater. They should also be capable of withstanding anticipated ground conditions, particularly where long-term monitoring is intended (see also Table A6.1).

For monitoring over short durations conventional materials (e.g. PVC) are usually satisfactory; under potentially aggressive conditions or for long-term installations, uPVC is preferable. Teflon^R (PTFE) or stainless steel are recommended in the US as materials least likely to cause significant error in groundwater quality monitoring due to chemical interaction with substances present in the water. Where composite constructions are used particular consideration must be given to jointing arrangements.

Teflon^R and stainless steel are more expensive than conventional materials but the additional cost of material should be weighed against potentially abortive design, implementation and monitoring costs if invalid data are collected.

A6.1.2.2 Casings and screens

Commercially manufactured casings and screens should be used to minimise construction variations between different monitoring installations. Screw threaded sections avoid the need to use (potentially contaminating) adhesives. The practice of forming slots on site by sawing is not acceptable since this results in variable screen performance (both within and between installations) and usually produces an ineffective screen, with slots too narrow to prevent clogging. Cutting on-site also exposes fresh PVC surfaces and increases the potential for leaching of organic contaminants (e.g. plasticisers, metal stabilisers).

Casing, screening and other materials should be carefully prepared before installation. Preparation includes thorough cleaning, and the protection of construction materials against contact with contaminated surface soils.

Packing materials (e.g. sand and gravel packs) should be consistent with the geological properties of the formation. Materials (e.g. quartz sand) imported on to the site for packing purposes should be free of contamination.

Table A6.1 *Possible construction materials*

Material	Comments
Fluoropolymer materials including:	
• polytetrafluoroethylene (PTFE) • tetrafluoroethylene (TFE) • fluorinated ethylene propylene (FEP) • perfluoroalkoxy (PFA) • polyvinylidene fluoride (PVDF)	Nearly totally resistant to chemical and biological attack, oxidation, weathering and ultraviolet radiation, have a broad useful temperature range (up to 550°C) Advantages include: almost completely inert to chemical attack including strong mineral acids and solvents, minimal sorption of chemical constituents from solutions (but some evidence for sorption of some halogenated compounds), minimal leaching of materials from the fluoropolymer structure Disadvantages include: high cost, handling difficulties (heavier, less rigid than PVC, slippery when wet), limited tensile strength of casing joints, low compressive strength, flexibility of pipestring[2]
Metallic materials including:	
• **carbon steel** • **low-carbon steel** • **galvanised steel**	Advantages include: generally stronger, more rigid and less temperature sensitive than plastic materials Note that carbon steel, low-carbon steel and galvanized steel are not recommended for use in most natural geochemical environments Disadvantages include the potential for corrosion (the difference between the corrosion resistance of carbon and low-carbon steels is negligible when buried in soils or in the saturated zone, galvanising does little to increase resistance). Corrosion products may affect water quality and corroded areas provide sites for chemical reactions and adsorption
• stainless steel (types 304 and 316)	Stainless steels perform well in most corrosive environments, particularly under oxidising conditions. There may be some susceptibility where there is microbial activity
Thermoplastic materials including:	
• polyvinyl chloride (PVC) • acrylonitrile butadiene styrene (ABS)	Generally weaker, less rigid and more temperature-sensitive than metallic casings, but adequate performance can be achieved in most applications by appropriate specification Advantages include: resistance to corrosion, lightweight, low cost, high strength to weight ratios, abrasion resistance, good durability under natural groundwater environments, flexible, workable and easily jointed PVCs may be adversely affected by; low molecular weight ketones, aldehydes and chlorinated solvents; creosote; petroleum distillates. Sorption of other compounds, e.g. chlorinated solvents, may also occur. When recharge rates are high purging may provide adequate protection against these effects
Fibre-reinforced materials including:	
• fibre-reinforced epoxy (FRE) types • fibre-reinforced plastic (FRP) types	Little practical experience. Chemical resistance and sorption properties will depend on the plastics matrix

The depth and location of the screen sections should be considered carefully, taking into account the requirements of the investigation, the permeability of the various geological strata and the nature of the contamination, if known. Whenever possible the screen should be fully submerged to avoid groundwater contact with the atmosphere where chemical reactions may occur. The potential for stratification of contaminants within the saturated zone should also be taken into account. For example, low density organic compounds will float on the groundwater surface and in this case screens should extend above the zone of saturation in order to detect such lighter substances.

A6.1.2.3 Annular seals

Sealing materials and methods of placement require careful consideration. Although bentonite is generally considered to offer good sealing properties, certain organic substances have been shown to migrate through bentonite with little or no attenuation. Neat Portland Cement pastes (possibly mixed with additives) may offer advantages over bentonite in certain applications.

A6.1.2.4 Surface finish

The surface finish of groundwater monitoring installations is important in protecting the installation from unauthorised access and accidental damage. In general, installations should be provided with a durable lockable cover. Measures taken to protect against vandalism (e.g. making the installation as inconspicuous as possible) may increase the risk of damage due to legitimate activities such as vehicle movements, ground maintenance activities etc. Consideration should be given to the use of physical protection measures (e.g. steel casing) to protect against accidental damage.

A6.1.2.5 Well development

It is crucial that all wells undergo a process of well development to ensure that they are functional at the time of installation, and capable of providing sediment-free samples for analysis. A wide variety of methods and techniques are available including bailing, surging and flushing with air and water. The basic principle is to create reverse flow into and out of the installation to break down deposits formed on the surface during installation, and to develop an appropriately graded filter pack within natural or imported layers adjacent to the installation.

Well development may have potential health and environmental impacts, particularly where air assisted methods are used. These are similar to those outlined above for drilling operations.

A6.2 SOIL VAPOUR ANALYSIS

Soil vapour/gas analysis is used to detect and measure volatile organic compounds (VOCs). It can be a valuable, powerful and economic tool in the investigation of some contaminated sites.

Sampling and measurement techniques may :

- permit immediate identification and/or measurement of the contaminants present, or
- integrate emissions over a period of time (often several weeks).

A range of different sampling point types may be used in the first (most common) application but shallow probes and various instrumental techniques are most often employed. Sampling devices can be placed at depths of 30 m or more using driven small diameter probes. Samples for analysis may be withdrawn under suction or by flushing the gas from a sample chamber into which it has entered through a gas-permeable membrane of high diffusion impedance[4].

Measurements are frequently made on-site although samples for off-site analysis can also be collected using either gas containers or absorption tubes. Substance-specific detection tubes may also be employed down exploratory holes.

When an integrated measurement is required, an adsorption device is placed in a hole in the ground. This is recovered after a suitable period of time, which may be several weeks, and the absorbed substances desorbed in the laboratory for analysis using gas chromatography, mass spectrometry and similar instrumental techniques. Adsorption may be onto propriety devices intended to extract gases from a gas flow, proprietary devices intended for other purposes (e.g., monitoring individual personnel exposures), or proprietary devices[5,6] specially developed for this purpose. In the last case, placement of the devices and measurement of the absorbed gases may be offered as a package deal by a specialist vendor.

The advantages of integrating techniques are that they can detect lower gas/vapour concentrations and they also smooth out variations in concentrations associated with fluctuating water tables and atmospheric pressure that would otherwise necessitate repeated and regular measurement.

Soil vapour measurements provide an indirect and relative measure of the concentration of volatile contaminants in the soil, either immediately surrounding the sampling point or at depth. When sampling is carried out on a regular grid, isopleths (lines of equal concentration) can be developed showing the magnitude and direction of movement of a pollution plume. These plots can then be used to help locate suitable positions for other types of sampling installations (e.g. deep probes or boreholes).

Although the measurements obtained may be only relative to the total concentrations of the substances in the soil or groundwater, they are likely to be of comparable reproducibility and reliability to measurements made on laboratory samples given the difficulty of obtaining representative samples and the potential for loss of volatiles during sampling and subsequent handling.

A6.3 GEOPHYSICAL TECHNIQUES

Geophysical investigations must be carried out as an integral part of the overall investigation process and all available information (i.e. the results of preliminary investigations and intrusive investigations to date) should be made available to the geophysicist[7,8]. This is essential to ensure that appropriate geophysical methods are selected and the investigation is properly designed. A list of available techniques is given in Box A6.3. A full review of their use is given by Reynolds and McCann[8].

> **Box A6.3** *Surface geophysical methods used in the investigation of contaminated sites*
>
> Seismic refraction
> Electrical resistivity
> Self potential including: ground conductivity, electromagnetic methods
> Ground probing radar
> Magnetic and magnetic gradiometry
> Gravity

Geophysical techniques can be employed to detect the boundary between strata having different physical properties. They measure differences in physical properties such as electrical conductivity (resistivity), bulk density (gravimetric), velocity of shock waves (seismic) or magnetic susceptibility. They can only detect a boundary where a distinct change in physical properties exists. Anomalies, such as near surface disturbances, may limit their usefulness.

Geophysical techniques may be of value in contamination investigations to:

- detect the boundaries between ground with different levels of contamination (where this affects bulk properties of the soil — which here includes any groundwater present)

- detect the boundaries between materials with different physical properties that incidently have different levels of contamination (e.g. the boundary between fill and natural ground or two types of fill)
- detect buried drums, tanks or other artifacts.

The first application is difficult but not impossible using certain recently developed techniques. The second may be practical depending on site conditions and in any case may form a useful part of the geotechnical investigation. The third may be possible on occasion.

The exact position of a boundary cannot always be identified because properties may change within a transition zone, rather than at a given point, or only over large distances where they are associated with migrating contaminants. The best results are obtained when ground conditions are uniform and simple, and there are large differences between the physical properties of the various formations.

The assessment of potential groundwater pollution associated with leachates from landfills etc. has been the subject of considerable research. In groundwater studies, electrical resistivity has been used widely because the flow of electrical current in the ground is largely a function of the electrical properties of the fluids present. Electromagnetic methods, such as ground conductivity measurements, are also applicable since the conductivity of leachate is usually significantly different to that of uncontaminated groundwater. The British Geological Survey has developed (and offers on a commercial basis) a multi-electrode resistivity system[9,10].

The application of geophysical techniques (e.g. ground penetrating radar) to archaeological investigations has met with considerable success. It is possible therefore, that where industrial sites of some complexity are being investigated, expertise in this subject area, as well as geotechnical engineering, may prove valuable.

A6.4 REMOTE SENSING

A variety of remote systems exist which may have limited application to the investigation of contaminated sites. They are more likely to be of value during the preliminary and exploratory phases of an investigation than the main or detailed phase. At best they give an indication of where to locate exploration holes and may assist in monitoring potentially hazardous situations on the ground e.g. hand-held or airborne thermal imaging equipment may indicate areas of underground combustion.

Hand-held equipment can detect temperature differences of as little as 0.1°C. Observations from airborne equipment can detect combustion at depths of tens of metres.

False colour infra-red photography may indicate areas of stressed vegetation caused, for example, by landfill gas or phytotoxic compounds in the soil (note stress may also be due to natural causes such as drought, flooding or disease). The technique requires expert interpretation but is comparatively cheap, especially when cameras are flown in model aircraft.

Some experimental work has been carried out using multi-spectral analysis[11]. This was successful in detecting differences between surface materials on the site investigated but requires specialist equipment for both data gathering and analysis. It is a potential tool for the future.

A6.5 FIELD EVALUATION OF AQUIFERS

An understanding of the response of an aquifer to abstraction is important in understanding groundwater hydrology (see standard texts for a description of the theory of aquifer behaviour[2,12]). To determine the yield of a groundwater system and to evaluate the movement of groundwater contaminants requires data on the:

- position and thickness of aquifers and confining beds
- transmissivity and storage coefficient of the aquifer
- hydraulic characteristics of the confining beds
- position and nature of the aquifer boundaries
- location and volumes of groundwater abstraction
- location, types and concentrations of contaminants.

Aquifer tests to determine temporal changes in water levels induced by abstraction are important tools in the determination of some of these parameters. Multiple well tests provide most information but useful data can be obtained from a single test well.

A6.6 GROUNDWATER MODELLING

Contaminant behaviour in groundwater is influenced by hydraulic gradients, geological factors, abstraction etc., in addition to the physical, chemical and biological processes which result in the dilution, attenuation and degradation of contaminants in the groundwater.

It is not generally feasible to provide sufficient monitoring points in the field to obtain a comprehensive analysis of groundwater conditions based on direct observation. Groundwater modelling techniques have been developed as a way of predicting contaminant behaviour based on fewer monitoring points and interpolation between points using mathematical formulae describing specific physical, chemical and biological processes.

Such models may also be used for risk assessment purposes (see Section 5) or to predict likely design requirements, efficiencies and duration of remedial operations in actual contaminated groundwater situations. In these applications, field data describing ground conditions, together with leaching/desorption data obtained through direct laboratory testing on collected samples, should be provided to inform the model.

Where laboratory and field based data are not available in sufficient detail, or in those cases where only an approximate indication of remedial requirements is needed, data from similar situations or applications may be used to predict expected outcomes. The greater the input of good quality monitoring data, the greater the accuracy of the result. Experience in the Netherlands suggests that for Dutch soils remediation duration can be predicted with an accuracy of only ± 50 %

The use of models for contaminant behaviour prediction, for risk assessment (see Section 5 and Appendix 10) or remediation purposes, requires the use of specialist expertise. In all cases, care should be exercised to ensure that the assumptions forming the basis of the model are appropriate and applicable to the situation under consideration.

A6.7 INVESTIGATION OF SPECIFIC MEDIA

A6.7.1 In-ground gases

Brief details only are provided here on the sampling of in-ground gases. Guidance on the investigation of sites for methane and other bulk gases (including advice on the selection of on-site and laboratory instrumentation) is provided in reference 13. A further CIRIA study is in progress to prepare more detailed guidance on the investigation of gassing sites[14].

In-ground gases may comprise:

- bulk gases such as oxygen, nitrogen, carbon dioxide, methane

- trace gases such as ethane, propane, helium, hydrogen sulphide, carbon monoxide, hydrogen (often inherently present but varying in proportion and concentration depending on source and therefore often a useful diagnostic tool — see reference 12)

- volatile organic compounds (VOCs).

Volatile organics may occur:

- as traces in landfill gas (either as an inherent component or as an indicator of the disposal of organic substances)

- arising from contamination of the ground with the substance in question or a complex mixture of substances (common materials such as petrol and fuel oils contain a variety of substances varying greatly in volatility).

The presence of VOCs in the latter case may be indicative of either a near surface spill or of pollution at considerable depth being transported on the surface of the groundwater table.

An indication of the presence and quantity of bulk gases can be obtained using portable instruments. These are easy to use and cost-effective for safety control purposes and for initial identification and location of the hazard, although their selection, use and interpretation require special care[13].

For accurate data on the concentration of any bulk gas and all trace gases, samples should be collected for laboratory analysis by gas chromatography/mass spectrometry (GC/MS). Samples should be collected using metal containers, pressurised if necessary (e.g. stainless steel 'Gresham™ Tubes'), which have been flushed through three times with the gas before final filling for analysis. Plastic or rubber containers are not recommended as they are relatively gas permeable and may absorb the gases. Pumps used to extract the sample should be of plastic or rubber composition when sampling at about 100 kPa, or of inert metal composition when higher pressures are involved.

When sampling in-ground gases, particular consideration should be given to the depth of sampling relative to the location of gas permeable strata, and to the need for long-term monitoring of the gas regime. The installation of permanent monitoring positions is invariably required with regular monitoring of in-situ concentrations and gas emission rates, and laboratory chromatography analysis of gas samples collected from each installation. In addition, ambient and in-ground temperature should be recorded on each monitoring occasion, together with atmospheric pressure, immediately before, during and after the collection of the sample, weather conditions including any recent rainfall, and depth to water table.

A6.7.2 Ambient atmospheres

Ambient atmospheres may be sampled for gases or dusts:

- to determine potential airborne hazards associated with a site

- as a health and safety measure to protect the workforce during site investigation or remediation or other works (e.g. decommissioning, decontamination or demolition)

- to protect public health or the environment in the vicinity of a contaminated site during site investigation, remediation or other works

- to provide confirmation that remedial action has been effective in reducing an airborne hazard.

General considerations and procedures have been described by Keith[15].

In short-term monitoring applications (e.g. health and safety during site investigation) both personal monitors (badge-type samplers) or absorption tubes can be used. The advantage of the first type of monitor is that it offers real-time exposure protection, indicating the point at which exposure is about to be, or has been, exceeded. Absorption tubes offer the advantage that a wider range of substances may be monitored, but detection requires laboratory analysis. Therefore, exposures can be assessed only after they have occurred. More detailed information on the use of such equipment for health and safety applications on contaminated sites is given in reference 16.

Hand held gas detection equipment, such as combustible gas detectors, oxygen depletion detectors and gas-specific absorption tubes are also invaluable tools in the detection and monitoring of potentially hazardous atmospheres[13,14].

A range of equipment is available for monitoring ambient air quality over longer time periods. Three basic types of equipment are available:

- extractive systems, in which measured volumes of air are drawn through a suitable filter or scrubbing medium which are periodically collected for analysis in the laboratory. Both particulate (e.g. metal-contaminated dust) and organic vapours can be measured.

- extractive systems which measure the contaminant directly in an extracted gas sample using spectrographic means

- optical systems which use spectrographic (e.g. infra-red, UV/visible) means to detect contaminants along a sampling path directed through the sampling area — some systems are capable of measuring several substances at ambient concentrations simultaneously. Equipment using laser sources is also under development for ambient air monitoring purposes.

The location of devices to monitor dust and gases emitted from the site requires careful consideration and should take into account the location of potentially sensitive targets relative to the anticipated movement of the potential hazards. Modelling techniques may be usefully employed to aid effective and economic location of sampling devices (see Appendix 10)

A6.7.3 Radioactivity

If the desk study indicates the potential presence of radioactive materials, on-site testing with appropriate precautions must be carried out as part of the investigation (see DD175:1988[17]). Most radioactive contamination on the surface of a site can be identified using portable instruments for the detection of alpha, beta, gamma and neutron emissions. Gamma emissions from buried substances can also be detected by such instruments. Other forms of buried radioactivity can only be identified from laboratory assay of samples. It is imperative that any investigations for radioactive substances are undertaken by appropriately qualified personnel and in strict accordance with the Ionizing Radiations Regulations 1985[18]. Appropriate precautions and arrangements are also required for any intended transport or disposal of such materials either as effluent or in bulk. This includes any radioactive material taken off-site for testing purposes. A good account of the investigation of a site containing radioactive materials has been provided by O'Brien et al[19].

A6.7.4 Drums

The sampling of drummed and other containerised materials is discussed briefly in Volume II (Decommissioning, decontamination and demolition). Issues to be considered include:

- staging
- identification
- opening
- health and safety procedures.

Prior to any sampling, an inventory of all the containers on the site should be completed. All information (capacity, labelling, condition) should be recorded and photographs taken. Depending on their location, position and condition, it may be necessary to place containers in an upright position and/or relocate them to a secure area prior to sampling. Drums containing liquid wastes can be under pressure or vacuum. A bulging drum should not be moved or sampled until the pressure can be safely relieved. Containers that can be moved should be positioned with the opening or bung upwards (if the integrity of the container will allow this). Containers should not be stacked.

The containers should then be marked with an identification number. Further photographs can usefully be taken at this stage.

The procedure used to open the container will depend on its condition and what is known about its contents. All containers should be opened with the utmost care. For drums, the bung opening should be loosened slowly with a non-sparking bung wrench. Remotely controlled opening or penetrating devices should be used when necessary. During container opening organic vapour concentrations should be monitored with portable instrumentation. Operations may usefully be recorded using televideo equipment. The condition of the drums may require them to be placed in a secure secondary container and the necessary materials should be to hand.

Containerised solid material (sludges, granules, powder) is usually sampled using a scoop or trowel, waste pile sampler, Veimeyer sampler/corer, sampling trier, or grain sampler. Liquid samples can be taken using open tube samplers, COLIWASA sampler, stratified sample thief, or a VACSAM sampler. Equipment should be carefully cleaned between sampling events.

Decisions will be required about whether to analyse each individual sample or to prepare composite samples. Detailed guidance and model protocols for sampling, compositing and compatibility testing of drum contents are available[20,21].

Guidance is also available on the procedures to be followed in the sampling of tanks[22], trucks, process vessels and similar large containers (see also Volume II).

A6.7.5 Flora and fauna

Sampling of flora and fauna may be required as part of the site investigation:

- to fully describe the distribution of the contaminant in the environment
- as part of the assessment of contaminant impacts on natural habitats
- as a means of monitoring contamination concentrations in targets over extended time periods.

At the simplest level, sampling may involve only the collection of samples of living material, such as plant foliage, or fruits, likely to be indicative of exposure to contaminants present in the ground, in waters or via atmospheric deposition.

A full ecological assessment (see Appendix 12) will involve the systematic sampling of various types of life forms, such as small mammals, fish life, mud-dwelling animals, invertebrates etc., requiring the use of standardised ecological sampling methods. This type of work should only be undertaken by those properly qualified to do so.

For simple applications, it is important to record:

- the type of sample (species, tissue type, gender where appropriate etc.)
- the location of the sample (cross referenced to relevant soil or water samples)
- the time of year (since uptake patterns in biological systems vary in response to seasonal factors)
- the field conditions at the time of sampling.

DD175:1988[17] provides guidance on sampling for flora and fauna and ISO is preparing test procedures to determine the effects of pollutants on selected flora and fauna (see Table A7.5 in Appendix 7). ADAS has provided guidance[23] on the sampling, preparation and analysis of agricultural crops. Guidance[24] from ADAS on sampling for nematodes and nematode cysts is indicative of the type of technique employed to recover microfauna.

Care should be taken not to destroy wild plants during sampling operations. Certain plants are protected by the Wildlife and Countryside Act 1981 (see Volume XII).

Already developed sites present particular difficulties and may require both sampling of existing vegetation and the planting of crops under controlled conditions for subsequent analysis (e.g. 'pot studies'). This may be done on-site or off-site.

The growth, or lack of growth, of test plants under controlled conditions may be used as an 'analytical' tool to assess the toxicity of soils to plants (phytotoxicity) and the potential for uptake of contaminants into human or other food chains. Some ISO standards are in preparation (see Table A7.5 in Appendix 7) and the ASTM has produced a wide range of tests (e.g. toxicity tests, sampling plankton, collecting benthic micro-invertebrates) for determining biological effects and the environmental fate of chemicals[25].

A6.7.6 Micro-organisms

Sampling soils and waters for micro-organisms may be required because:

- they may be harmful and present a risk of infection to workers or to site users or occupiers
- they provide a measure of 'soil quality' and any adverse impacts that may have been associated with contamination

- if suitably stimulated, they may be used to destroy contaminants (see biological treatment in Volumes VII and IX).

Brief guidance on sampling for micro-organisms is given in DD175:1988[17] and ISO DIS 10381-6[26].

Special techniques are required to minimise accidental contamination of the sample and to obtain representative material. Good microbiological practice (e.g. sterile techniques) should be used to prevent cross-contamination and to minimise any risk of infection. It is particularly important to handle and store samples correctly and to label and contain samples so as to avoid danger of unwitting contact with pathogens.

A6.7.7 Metallurgical slags

When tests are required to determine the volume stability of blastfurnace and steelmaking slags it is customary to take samples of about 25 to 50 kg. These are then inspected in the laboratory to select suitable sub-samples for testing. The guidance of the laboratory carrying out the testing should be obtained, and it may be appropriate for someone experienced in the assessment of slags to supervise the sampling. The testing of slags is discussed in Appendix 7.

REFERENCES

1. INTERNATIONAL ORGANISATION FOR STANDARDISATION. *Guidance on sampling techniques: Soil quality – Sampling: Part 2.* CD 10381 Part 2. Committee Draft (April 1993)

2. UNITED STATES ENVIRONMENTAL PROTECTION AGENCY. *Handbook of suggested practices for the design and installation of ground-water monitoring wells.* USEPA (Washington DC), 1991, EPA/600/4-89/034

3. NAYLOR, J.A., ROWLANDS, C.D. and BARBER, C. *The investigation of landfill sites.* Technical Report TR 91. Water Research Centre (Medmenham), 1978

4. ROBITAILLE, G.E. Quantitative in situ soil gas sampling. In: *Proceedings of the Eighth Annual Waste Testing and Quality Assurance Symposium.* American Chemical Society 1992, pp 2-14

5. VIELLENAVE J H and HICKEY J C. Use of high resolution passive soil gas analysis to characterize sites contaminated with unknowns, complex mixtures, and semivolatile organic compounds. *Hazardous Materials Control*, 1991, $\underline{4}$(4), 42-49.

6. ANON. Advanced soil gas sampling maps hydrocarbon in contaminated land. *Industrial Waste Management.* 1994 (May) pp 186-70

7. DARRACOTT, B.W. and McCANN, D.M. Planning engineering geophysical surveys. In: *Geological Society Special Publication No. 2.* Geological Society (London), 1986, pp 85 – 90

8. REYNOLDS, J.M. and McCANN, D.M. Geophysical methods for the assessment of landfill and waste disposal sites. In: *Proceedings of the Second International Conference on Construction on Polluted and Marginal Ground.* Forde, M.C. (ed.) Engineering Technic Press (Edinburgh) 1992.

9. JACKSON, P.D., MELDRUM, P. and WILLIAMS, G.M. *Principles of a computer controlled multi-electrode resistivity system for automatic data acquisition.* Report WE/89/32. British Geological Survey (Keyworth), 1989

10. WILLIAMS, G.M. and JACKSON, P.D. *A multi-electrode system for hydraulic characterisation.* Paper to Hydrogeological Group Meeting of the Geological Society, 'Technical Advances in Downhole Investigation,' 1990

11. BRIDGES, E.M. *The use of remote sensing in the identification, mapping and monitoring of contaminated land.* University of Swansea (Swansea), 1984

12. FREEZE, R.A. and CHERRY, J.A. *Groundwater.* Prentice Hall (Englewood Cliffs, New Jersey), 1979

13. CROWHURST D., and MANCHESTER S.J. *The measurement of methane and other gases from the ground.* Report 131, CIRIA (London), 1993

14. *Methane investigation strategies.* FR/CP/14. CIRIA (London), 1993.

15. KEITH, L.H. *Environmental sampling and analysis: A practical guide.* Lewis Publishers (Chelsea MI), 1991

16. STEEDS J., SHEPHERD E. and BARRY D.L. *A guide to safe working practices for contaminated sites.* Report 132 CIRIA (London), (in the press).

17. BRITISH STANDARDS INSTITUTION. *Draft for Development Code of Practice for the identification of contaminated land and for its investigation.* BS DD175:1988. BSI (London), 1988

18. *Ionizing Radiation Regulations 1985.* HMSO (London), 1985

19. O'BRIEN, A.A., STEEDS, J.E. and LAW, G.A. Case study: Investigation of the Long Cross and Barracks Lane landfill sites. In: *Proceedings of a Conference on Planning and Engineering of Landfills.* West Midlands Geotechnical Society (Birmingham), 1992, pp 31–34

20. UNITED STATES ENVIRONMENTAL PROTECTION AGENCY. *Guidance document for cleanup of surface tank and drum sites.* USEPA (Washington DC), 1985, OSWER Directive 9380.0-3.

21. MONTGOMERY, R.E., REMETA, D.P. and GRUENFELD, M. Rapid on-site methods of chemical analysis. In: *Contaminated Land: Reclamation and Treatment.* Smith M. A. (ed.) Plenum (London), 1985, pp 257-310

22. UNITED STATES ENVIRONMENTAL PROTECTION AGENCY. *State-of-the-art procedures and equipment for internal inspection of underground storage tanks.* USEPA (Washington DC), 1991, EPA/600/2-90/061

23. MINISTRY OF AGRICULTURE FISHERIES AND FOOD. *Laboratory methods for work with plant and soil nematodes.* MAFF Reference Book 402. HMSO (London), 1986

24. MINISTRY OF AGRICULTURE FISHERIES AND FOOD. *Sampling of soils, soilless growing media, crop plants and miscellaneous substances for chemical analysis.* MAFF (London), 1979

25. AMERICAN SOCIETY FOR TESTING AND MATERIALS. *Annual Book of ASTM Standards, Section 11: Water and environmental technology.* ASTM (Philadelphia) annual publication.

26. BRITISH STANDARDS INSTITUTION/INTERNATIONAL ORGANISATION FOR STANDARDISATION. ISO DIS 10381-6: *Soil Quality – Sampling: Part 6: Guidance on the collection, handling and storage of soil for aerobic microbial processes in the laboratory.* Available from BSI(London), 1993

Further reading

McCANN, D.M. Geophysical methods for the assessment of landfills and waste disposal sites: a review. *Land Contamination and Reclamation,* 1994, **2** (2), 73-83

WATSON, C. (ed.) *Official and standardized methods of analysis* (3rd edition). Royal Society of Chemistry (London), 1994

Appendix 7 Analytical and testing strategies and methods

This Appendix provides general guidance on analytical and testing strategies. It is not intended to be fully comprehensive. For detailed guidance reference should be made to the various publications cited and to other standard texts.

A7.1 AVAILABLE GUIDANCE

A7.1.1 General availability

In the UK few standardised analytical methods have been specifically developed for the analysis of contaminated soils although methods developed for other applications may be adaptable for the purpose. There is an extensive range of 'methods for the examination of water and associated materials' ('blue books') produced by the UK Standing Committee of Analysts (SCA)[1]. The SCA is a joint body of the Department of the Environment and the water industry. The Ministry of Agriculture Fisheries and Food also publishes analytical methods for agricultural materials including soils and plant materials[2], some of which may be appropriate to the assessment of contaminated sites (but see comments in Section A7.1.2). Some major UK organisations (e.g. British Gas) require that specific reference analytical methods are employed by contract laboratories.

Analytical methods for contaminated soils and measurement of a range of environmentally related parameters have been produced in other countries including Germany, the Netherlands and the USA. In Canada, Environment Canada has published[3,4] guidance on appropriate analytical methods to be used in connection with the National Contaminated Site Remediation Program (see Volume XII). These methods are drawn from a variety of sources. In addition, Technical Committee 190 (Soil Quality) of the International Organisation for Standardisation (ISO) is currently working on a 'soil quality' programme covering both contaminated and uncontaminated soils. The ISO programme includes the development of standards for:

- terminology and vocabularies for describing soils and sites
- codification of site data to aid computerisation and data exchange
- sampling
- chemical analytical methods
- determination of physical properties (laboratory and in-situ methods)
- biological assessment methods.

The Comité European de Normalisation (CEN) is considering whether it also has a role to play in this subject area. CEN committee TC 223 is producing a range of standard analytical methods for soil improvers and growing media. TC 292 is concerned with characterisation of waste products. CEN standards are important because they are frequently specified in European Community Directives when either chemical or some other form of analysis is required and are automatically adopted as British Standards.

The ISO also has an active programme to develop test methods for water and sediments etc. (Technical Committee 147). ISO methods commonly form the basis of national standards including those issued in the UK by the Standing Committee of Analysts.

A policy decision has been made by the British Standards Institution (BSI) to devote the limited resources available both within BSI and member organisations to the ISO work on soil quality, rather than to the development of national standards. The ISO standards will be adopted as British Standards as they emerge and are therefore of great importance.

Note that the initial set of ISO analytical methods for soil will generally be reference methods, i.e. methods against which laboratories may judge the performance of any in-house or national standard they employ.

Table A7.1 lists major organisations producing relevant standards.

Table A7.1 *Organisations producing standard analytical and test methods for soils and waters*

Organisation	ISO	CEN	BSI	SCA	MAFF	NNI	DIN	ASTM	USEPA	AHPA
Soils (analysis)	YES	/	ISO	NO	YES	YES	YES	YES	YES	NO
Water (analysis)	YES	?	YES	YES	NO	YES	YES	YES	YES	YES
Soils (biological)	YES	/	ISO	NO	YES	?	YES	YES	?	NO
Soils (physical)	YES	/	ISO	NO	YES	YES	YES	YES	YES	NO

Abbreviations:

ISO   International Organisation for Standardisation
CEN   Comité European de Normalisation
BSI   British Standards Institution
SCA   Standing Committee of Analysts
MAFF Ministry of Agriculture Fisheries and Food
NNI   Nederlands Normalisatie-Instituut
DIN   Deutsches Institut für Normung e.V.
ASTM American Society for Testing of Materials
USEPA United States Environmental Protection Agency
AHPA American Public Health Association

General notes:

1. BSI, NNI, DIN, and ASTM also provide standard methods for testing soils for engineering purposes.

2. Other national standardisation bodies (e.g. UNICHIM in Italy)[5,6] also provide limited ranges of standard analytical methods for soils.

3. CEN methods for soil improvers and growing media, and for characterisation of waste products may have limited applicability.

4. The American Public Health Association, Water Environment Federation and American Water Works Association, jointly publish 'Standard Methods for the Examination of Water and Wastewater' (18th Edition published 1993 by WEF (Alexandria, Va)).

A7.1.2 United Kingdom Standards

Standardised analytical methods are issued in the UK by the :

- British Standards Institution

- Department of the Environment Standing Committee of Analysts
- Ministry of Agriculture Fisheries and Food

British Standard (BS) geotechnical methods are contained in BS 1377:1990. There are currently few BS analytical methods specific to agricultural aspects or to soil contamination but, ISO methods will be adopted as they appear.

No methods are specified for use in connection with the Interdepartmental Committee on the Redevelopment of Contaminated Land (ICRCL) 'trigger' values[7,8] (see Section 6 for a discussion of these) with the exception of PAHs and acid soluble chromium for which a method is described in broad terms – no detailed method has been published. However, the Department of the Environment commissioned a study in 1993 to identify available analytical methods which might be suitable for use with the 'trigger' values.

The Standing Committee of Analysts has produced a wide range of methods for the examination of 'waters and associated materials'. These methods are frequently issued in parallel as British Standards and are often closely related to, or coincide with, ISO methods.

The Ministry of Agriculture Fisheries and Food (MAFF) publishes methods for the analysis of agricultural materials including the preparation of soil and plant materials; trace elements in soils and plants; and nutrient status of soils[2]. These and other MAFF approved methods are kept under review and are the subject of continuing programmes of inter-laboratory testing. Note that the soil preparation methods described (e.g. grind to <2 mm) are specific to agricultural situations: they are not appropriate for contamination investigations where human health and environmental impacts are being assessed.

The Water Research Centre, working for the National Rivers Authority, has reviewed leaching tests for contaminated soils with a view to establishing a standard method to assess the potential of contaminated site redevelopment activities to pollute surface and groundwater. Following this work, the NRA has published[9] a recommended leach test which was still being evaluated in inter-laboratory trials in January 1995.

A7.1.3 ISO Standards

ISO standards relating to soil quality, with an indication of their current (May 1994) status are listed in Tables A7.2 to A7.6 (see Box A7.1 for key to Tables). Those published as British Standards by January 1995 are listed in Table A7.7.

Table A7.2 *ISO standards relating to soil quality – terminology*

Status/ref.	Title
CD 11259	Soil quality – description of soils and sites
DIS 11074-1	Soil quality – Vocabulary – Part I: terms and definitions relating to the protection and pollution of soil
CD 11074-2	Soil quality – Vocabulary – Part II: terms and definitions relating to sampling
WI 11074-3	Soil quality – Vocabulary – Part III: terms and definitions relating to risk assessment and soil rehabilitation
WI 11259	Soil quality – Codification of soil agronomic and analytical data

Box A7.1 *Key to Tables A7.2 to A7.6*

ISO = International Standard	Will be adopted by BSI as British Standard
ISO/TR = Technical Report	Published by BSI as Draft for Development
DIS = Draft International Standard	Distributed for formal comment and approval by all ISO members (circulated widely in member countries)
CD = committee draft	Distributed for comment to ISO members active in work of TC 190
WD = working draft	Committee paper
WI = work item	
NP = new project	

Table A7.3 *ISO standards for soils – sampling*

Status/ref.	Title
CD 10381-1	Part 1: Guidance on the design of sampling programmes
CD 10381-2	Part 2: Guidance on sampling techniques
CD 10381-3	Part 3: Guidance on safety
CD 10381-4	Part 4: Guidance on the procedure for the investigation of natural, near-natural and cultivated sites.
CD 10381-5	Part 5: Guidance on procedures for the investigation of soil contamination of urban and industrial sites
ISO 10381-6	Part 6: Guidance on the collection, handling and storage of soil for the assessment of aerobic microbial processes in the laboratory (BS 7755: Section 2.6: 1994)

Table A7.4 *ISO methods for soil – chemical analysis*

Status/ref.	Title
CD 10382	Determination of organochlorine pesticides and polychlorinated biphenyls
DIS 10390	Determination of pH
DIS 10693	Determination of carbonate content – volumetric method
DIS 10694	Determination of organic and total carbon after dry combustion
ISO/TR 11046	Method for determination of mineral oil content – method by infra-red spectrometry and gas-chromatograph method (BSI/DD220:1994)
CD 11047.2	Determination of cadmium, chromium, cobalt, copper, lead, manganese, and zinc - flame and electrothermal atomic absorption spectrophometric method
DIS 11048	Determination of water soluble and acid soluble sulfate
DIS 11260	Determination of cation exchange capacity and base saturation - Method using barium chloride solution
DIS 11261	Determination of total nitrogen – Kjeldahl method using titanium dioxide as catalyst
CD 11262	Determination of cyanide
ISO 11263	Determination of phosphorus – Spectrometric determination of phosphorus soluble in sodium hydrogen carbonate solution (BS 7755: Section 3.6:1995)
WD 11264	Determination of persistent herbicides
DIS 11265	Determination of specific electrical conductivity
DIS 11464	Pretreatment of samples for physico-chemical analysis
ISO 11465	Determination of the dry matter and water content on a mass basis by a gravimetric method (BS 7755: Section 3.1: 1994)
DIS 11466	Extraction of trace metals soluble in aqua regia
CD 13536	Determination of potential cation exchange capacity and base saturation – Method according to Bascomb at pH 8.1
CD 13877	Determination of polynuclear aromatic hydrocarbons – HPLC method
CD 13878	Determination of total nitrogen content after dry combustion ('element analysis')
NP 14154	Determination of phenols and chlorophenols
CD 14235	Determination of organic carbon by sulfochromic method
NP 14254	Determination of exchangeable acidity
CD 14255	Determination of soluble nitrogen fractions
CD 14256	Determination of inorganic forms of nitrogen in field-moist soils - method using 1 mole/litre potessium chloride solution as extractant
CD 14507	Pretreatment of samples for determination of organic contaminants

Table A7.5 *ISO methods for soils – biological assessment*

Status/ref.	Title (all have style: Soil quality – part 4: Biological methods : Section 4x)
DIS 11266-1	Guidance on laboratory testing for biodegradability of organic chrmicals in soil — under aerobic conditions
CD 11266-2	Guidance on laboratory testing for biodegradability of organic chemicals in soil — under anaerobic conditions
NP 11267	Determination of effects on soil fauna - Inhibition of reproduction of *Collembela* by pollutants
ISO 11268-1	Effects of pollutants on soil fauna - — Determination of acute toxicity to earthworms (*Eisenia fetida*) using artificial soil substrate (BS 7755: Subsection 4.2.1: 1994)
DIS 11268-2	Effects of pollutants on soil fauna — Determination of inhibition of reproduction of earthworms (*Eisenia fetida*)
NP 11268-3	Effects of pollutants on soil fauna - Guidance on field testing on earthworms (*Eisenia fetida*)
ISO 11269-1	Effects of pollutants on soil flora — Method for the measurement of inhibition of root growth (BS 7755: Subsection 4.3.1: 1994)
DIS 11269-2	Determination of the effects of pollutants on soil flora — Measurement of the effects of chemicals on emergence and growth of higher plants (to be BS 7755: Subsection 4.3.2)
DIS 11269-3	Determination of the effects of pollutants on soil flora — Method for measurement of germination
NP 11269-4	Determination of the effects of pollutants on soil flora - Part 4: Guidance on the field testing of higher plants
CD 14238	Determination of nitrogen mineralisation in soils ans influence of chemicals on the process
CD 14239	Laboratory testing for for mineralisation of organic chemicals in soils under aerobic conditions
CD 14240-1	Determination of bio-mass - Part 1: Respiration method
CD 14240-2	Determination of bio-mass - Part 2: Fumigation extraction method

A7.2 ANALYSIS OF SPECIFIC MEDIA

A7.2.1 Soils/fills

Sampling methods, handling and preparation are very important and failure to observe proper procedures can have a profound affect on the value of subsequent analyses. The choice of pre-analytical procedures (sub-sampling, grinding etc.) and analytical methods have been discussed by Lord[10] and others[11-14]. Detailed descriptions of non-standardised methods can be found in the scientific literature including conference proceedings (e.g. reference 15).

Sample pre-treatment requirements in any standard analytical method should always be observed unless there are compelling reasons to deviate, in which case, the reasons for the deviation should be reported.

Standard sample pre-treatment methods for soils contaminated with both inorganic and organic contaminants have been prepared by ISO (see Table A7.3). Pre-treatment methods are also being developed for physical testing and for biological testing, where appropriate.

Table A7.6 *ISO methods for soils — physical characteristics*

Status/ref.	Title (all have style : Soil quality — part 5: Physical methods: section 5x)
DIS 10573	Determination of the water content of the unsaturated zone — Neutron depth probe method
DIS 11508	Determination of particle density
CD 11271	Determination of redox voltage — field method
DIS 11272	Determination of dry bulk density
CD 11273-1	Determination of aggregate stability — Part 1: Tensile strength measurement (crushing strength)
CD 11273-2	Determination of aggregate stability — Part 2: Method by shear test
DIS 11274	Determination of water retention characteristic — laboratory methods
DIS 11275-1	Determination of hydraulic conductivity
CD 11275-2	Determination of the unsaturated hydraulic conductivity and water retention characteristics - Wind's evaporation method
DIS 11276	Method of determination of pressure potential — Tensiometer method
DIS 11277	Determination of particle size distribution in mineral soil material — method by sieving and sedimentation following removal of soluble salts, organic matter and carbonate
DIS 11461	Determination of soil water content on a volume basis - gravimetric method
CD 12229	Determination of water content — Time domain reflectometry (TDR) method

Table A7.7 *ISO methods for soils available as British Standards (January 1995)*

Status/ref.	Title
BS 7755: Section 2.6: 1994	Soil Quality: Part 2 Sampling: Section 2.6 Guidance on the collection, handling and storage of soil for the assessment of aerobic microbial processes
BS 7755: Section 3.1: 1994	Soil Quality: Part 3 chemical methods: Section 3.1 Determination of the dry matter and water content on a mass basis by a gravimetric method
BS 7755: Section 3.6: 1995	Determination of phosphorus: Spectrometric determination of phosphorus soluble in sodium hydrogen carbonate solution
BS 7755: Subsection 4.2.1: 1994	Soil Quality: Part 4 Biological methods: Section 4.2 Effects of pollutants on fauna: Subsection 4.2.1 Determination of acute toxicity to earth worms (*Eisenia fetida*) using artificial soil substrate
BS 7755: Subsection 4.3.1: 1994	Soil Quality: Part 4 Biological methods: Section 4.3 Effects of pollutants on flora: Subsection 4.3.1 Methods for the measurement of inhibition of root growth
DD 220: 1994	Soil quality — Determination of mineral oil content — Method by infra-red spectrometry and gas chromatographic method

Key factors to be taken into account include:

- sample pre-treatment must not result in loss of contaminants through volatilisation or microbial activity, or result in change in chemical composition (e.g. due to oxidation or reduction)

- subject to the above considerations, and the specific nature of the tests, samples should be ground finely (usually to less than 250 μm) to ensure analytical reproducibility and accuracy

- analyses may be made on 'as received'; on air-dried (not greater than 40°C) or on chemically dried samples. They should not be made on oven-dried (e.g. >110°C) samples.

A7.2.2 Water : on-site measurements

On-site measurements may be made using instrumental techniques either directly in the body of water (river, down hole in borehole) or in an extracted sample. They are intended to provide a direct and rapid result for those parameters that are unstable during storage.

A7.2.3 Water : laboratory measurements

A wide range of standard methods are available (see also reference 15).

A7.2.4 Non-aqueous phases

Non-aqueous phases require careful handling to avoid loss of volatiles and disproportionation (i.e. preferential loss of some components) of the sample etc. Special preparation may be required prior to analysis using instrumental techniques such as infra-red analysis and gas chromatography/mass spectrometry. Sample preparation is often the key issue rather than the analytical method itself.

A7.2.5 Toxicity and genotoxic testing

A discussion of genotoxicity testing can be found in Simmons[12]. The US Environmental Protection Agency has published procedures for short term genotoxic testing (e.g. for carcinogens and mutagens)[16].

A7.2.6 Gas: on-site measurements

A detailed review of on-site methods of analysis for methane, carbon dioxide and oxygen is provided in a recent CIRIA report[17].

Available methods for the detection of volatile organic compounds include instrumental techniques and the use of detector tubes. Instrumental methods include portable gas chromatographs, infrared analysers etc. Detector tubes are available for wide range of gases but are of limited sensitivity. These may be used down exploratory holes.

A7.2.7 Gas : laboratory measurements

Laboratory methods for gas analysis are reviewed in a recent CIRIA report[17]. The principal methods used are gas chromatography (see Box A7.2) and mass spectrometry (see Box A7.3). Modern laboratory-based gas chromatographs are able to determine compounds in the concentration range 0.1 part per billion to 1 part per million. The operation of gas chromatographs requires specialist knowledge in the choice of separating columns, detectors and operation. Gas chromatographs operate by separating the components of the gas mixture. They are often used in conjunction with mass spectrometers which can identify individual compounds.

A7.2.8 Macro flora and fauna

ISO methods are being developed to determine the effects of pollutants on soil macro flora and fauna (see Table A7.5). MAFF methods[2] are applicable for some plant and animal material.

> **Box A7.2** *Gas chromatography*
>
> Gas chromatographs separate gases and vapours by passing them through a separation column in a stream of carrier gas (usually an inert gas such as helium, argon or nitrogen). The degree of separation of the mixture depends on the length of the column and its affinity to various components of the gas mixture. A wide range of different types of columns are available. Selecting the most appropriate therefore requires expertise and knowledge of both the types of column needed to separate particular gases or vapours, and the types of components that may be present in the sample.
>
> Separated gases are measured in a detector. The choice of detector depends on the nature of the gases being measured.

> **Box A7.3** *Mass spectrometry*
>
> Mass spectrometers are used to identify and measure chemicals by determining their mass (molecular weight) and 'mass spectra'. Each molecule has an essentially unique mass spectrum. It may be extremely complex and analysis is usually carried out by comparing the sample spectrum with a library of reference spectra using computers. Specialist expertise is therefore required in both the operation of the instrument and interpretation of the results.

A7.2.9 Radioactivity

The ISO committee on soil quality (TC 190) had a number of work items addressing contamination with radioactive nuclides. This work is at present in abeyance due to lack of interest.

A7.2.10 Asbestos

Methods for determination of asbestos in soil are being developed in the UK and consideration is being given (August 1994) to production of a British Standard. A proposal to ISO TC 190 to produce an International Standard was not accepted.

A7.2.11 Volume stability of blastfurnace and steelmaking slags

Blastfurnace and steel making slags have markedly different compositions (information on typical compositions of modern slags have been provided by Barry[18]). 'Typical' compositions have also tended to change as parent metallurgical processes have been modernised. Because of their chemical or mineralogical composition, some slags may undergo expansive reactions in the presence of water, or be otherwise unstable. Steelmaking slags pose the greatest problem but there is some evidence that some old blastfurnace slags, which differ in composition from modern slags, may also be expansive. Typically mixtures of slag types will be encountered on sites, often in association with other wastes, such as flue dusts and spent refractories. Some of the latter may also be reactive/expansive.

Blastfurnace slags have been used for many years as construction aggregates (e.g. roadstone, road base, railway ballast, in asphalt, concrete aggregate), and in the production of cement. As a result, a range of tests (e.g. BS 1047[19]) have been developed in the UK and other countries to confirm the suitability of blastfurnace slag for the intended use, including avoiding the dangers of volume instability.

Steelmaking slags, although used as aggregate in limited applications (e.g. roadstone for surface dressing), have not been the subject of such standardisation in the UK, although standardised test methods have been developed in a number of countries including the USA (ASTM D4792[20]), Germany and Belgium for assessing potential volume instability.

British Standard methods developed for assessing the suitability of blastfurnace slag for use as aggregate are not generally applicable to the assessment of steelmaking slags; the chemical composition and hence possible mechanisms of reaction are too different.

There are six commonly recognised chemical reactions leading to volume instability in blastfurnace or steel making slags (see Boxes A7.4 and A7.5).

- free lime hydration – typical of steel slags but may occur in very old blast furnace slags
- periclase hydration – typical of steel slags, occasionally occurs in blast furnace slags
- sulphoaluminate formation – primary reaction of some blast furnace slags but secondary product in steel slags
- iron unsoundness – affects blast furnace slags
- rusting of metal inclusions – characteristics of steel slags
- 'lime unsoundness' – characteristic of blast furnace slags.

There have also been recent suggestions that thaumasite formation might lead to instability in some old blastfurnace slags.

Box A7.4 *Main reactions leading to instability of steel making slags*

Free lime hydration: Calcium oxide (free lime) found in steel making slags arises from the limestone (calcium carbonate) added during the steel making process. It hydrates to form calcium hydroxide, and subsequently carbonates, with a large (100%) volume increase. It may also react with any sulphate present to form calcium sulphate (gypsum) which can then further react to form calcium sulphoaluminate.

Periclase hydration: Magnesium oxide (periclase), arising from magnesian limestone or dolomite, may hydrate to form brucite with a large volume increase (130%). However the reaction is generally very slow.

Rusting: Rusting of the free metal frequently present in steelmaking slags is due to one of a range of hydrolysis reactions covered by the blanket term of 'iron unsoundness'.

Any test programme relating to slag stability should include:

- major oxide analysis (CaO, MgO, SiO_2, Al_2O_3, FeO/Fe_2O_3)
- X-ray analysis and/or optical microscopy to determine mineralogy
- determination of free lime content
- determination of sulphate, sulphide etc.
- appropriate volume stability tests.

Major oxide analysis allows the type of slag to be identified and, if it falls within the range of typical blastfurnace slags, a calculation to be made (according to BS 1047)[19] of the possible presence of periclase (MgO) or dicalcium silicate ($2CaO.SiO_2$) in the slag.

> **Box A7.5** *Main reactions leading to instability of blastfurnace slags*
>
> **Periclase hydration:** Magnesium oxide (periclase), occasionally present in blastfurnace slags, may hydrate to form brucite, also with a large volume increase (130%). Equations and tests included in BS 1047[19] are intended to exclude blastfurnace slags which might contain periclase from use as aggregate.
>
> **Iron unsoundness:** This is properly the term used to describe a specific hydrolysis reaction involving an iron sulphide likely to occur in blastfurnace slag. However, it is conveniently used to describe hydrolysis of any iron sulphide, metal or low oxidation state oxide. Such hydrolysis reactions commonly require acid conditions.
>
> **Sulphoaluminate formation:** Blastfurnace slag glass can react with sulphate under alkaline conditions to form sulphoaluminates. The volume expansion is about 130% relative to the glass.
>
> **'Lime unsoundness':** This is the misleading term given to the expansion (leading to a 10% volume increase) resulting from the change (inversion) of higher temperature isomorphs of dicalcium silicate to the gamma form stable at normal temperatures. The reaction, which typically occurs during cooling of the slag, is capable of reducing hard crystalline blastfurnace slag to a fine powder (as process known as 'falling' or 'dusting'. Equations and microscopic tests included in BS 1047 are intended to exclude slags which might contain this compound from use as aggregate.

Mineralogical analysis is used to determine which minerals are present, thus confirming the type and condition (e.g. crystalline or glassy/vitrified) of the slag. It may also identify known reactive (CaO, MgO) or unstable (dicalcium silicate) compounds, or the products of reaction (e.g. calcium sulphoaluminates).

Determination of free (uncombined) lime can provide a direct measure of potential reactivity.

Determination of sulphide and sulphate are relevant to both stability and pollution potential.

Iron unsoundness in blastfurnace slags may be determined using the method specified in BS 1047. This can also be adapted to test for similar reactions in steel making slags.

Potential instability due to free lime, free periclase and sulphoaluminate reactions is usually determined in a test in which compacted and confined samples of slag are heated in water to elevated temperatures for a period of time (from several days to several weeks)[21]. This is not a standardised method in the UK and care should be exercised to ensure that the laboratory carrying out the work is fully familiar with slag properties and the purpose and nature of the test. It requires a significant amount of judgement in the selection of the test samples from the large volumes brought from sites. The version of the test has been standardised by the American Society for Testing and Materials (ASTM)[20].

A7.3 QUALITY ASSURANCE DURING ANALYSIS

Quality assurance (QA) in respect of chemical and other forms of analysis defines the acceptance limits of measurement and monitoring data in terms of sensitivity, reproducibility, detection limits and accuracy. This subject is dealt with in Section 3. General discussion of quality in respect of environmental data can also be found in references 13 and 14.

REFERENCES

1. STANDING COMMITTEE OF ANALYSTS. *Methods for the examination of waters and associated materials — various titles.* HMSO (London), various dates

2. MINISTRY OF AGRICULTURE, FISHERIES AND FOOD. *The analysis of agricultural materials.* Reference Book 427. HMSO (London), 1985

3. CANADIAN COUNCIL OF MINISTERS OF THE ENVIRONMENT. *Guidance Manual on sampling, analysis and data management for contaminated sites, Volume 1: Main Report*, CCME (Winnipeg), 1993. Report CCME NCS 62E

4. CANADIAN COUNCIL OF MINISTERS OF THE ENVIRONMENT *Guidance Manual on sampling, analysis and data management for contaminated sites. Volume 2: Analytical method summaries* CCME (Winnipeg) 1993, Report CCME NCS 66E

5. UNICHIM. *Soil analysis, Part I: Manual methods.* UNICHIM (Italian Association for Standardisation in the Chemical Industry) (Milan) 1991. Manuele N.145 (1991) English Edition.

6. UNICHIM. *Soil analysis, Part II: Semiautomatic methods.* UNICHIM (Italian Association for Standardisation in the Chemical Industry) (Milan) 1991. Manuele N.145 (1991) English Edition.

7. INTERDEPARTMENTAL COMMITTEE ON THE REDEVELOPMENT OF CONTAMINATED LAND. *Guidance on the assessment and redevelopment of contaminated land.* ICRCL 59/83 (2nd edition). DOE (London), 1987

8. INTERDEPARTMENTAL COMMITTEE ON THE REDEVELOPMENT OF CONTAMINATED LAND. *Notes on the restoration and after care of metalliferous mining sites for pasture and grazing.* ICRCL 70/90. Department of the Environment (London), 1990

9. NATIONAL RIVERS AUTHORITY. *Leaching tests for assessment of contaminated land: Interim NRA guidance,* R& D Note 301. NRA (Bristol), 1994

10. LORD, D.W. Appropriate site investigations. In: *Reclaiming Contaminated Land.* Cairney, T. (ed.) Blackie (Glasgow), 1987

11. MONTGOMERY, R.E., REMETA, D.P. and GRUENFELD, M. Rapid on-site methods of chemical analysis. In: *Contaminated Land: Reclamation and Treatment.* Smith, M.A. (ed.) Plenum (London) 1985, pp 257-310.

12. SIMMONS, S.M. (ed.) *Hazardous Waste Measurements.* Lewis Publishers (Chelsea MI), 1990

13. KEITH, L.H. *Environmental sampling and analysis: a practical guide.* Lewis Publishers (Chelsea MI), 1991

14. DAVIES, B.E. (ed.) *Applied soil trace elements.* Wiley (London), 1980

15. AMERICAN CHEMICAL SOCIETY. *Proceedings 8th Annual Waste Testing and Quality Assurance Symposium, Arlington VA, 1992.* ACS, 1992

16. UNITED STATES ENVIRONMENTAL PROTECTION AGENCY. *Environmental assessment: Short-term tests for carcinogens, mutagens and other genotoxic agents.* USEPA Health Effects Research Laboratory (Research Triangle Park NC) 1979, EPA/625/9-79/003

17. CROWHURST, D. and MANCHESTER, S.J. *The measurement of methane and other gases from the ground.* Report 131. CIRIA (London), 1993

18. BARRY, D.L. Former iron and steelmaking plants. In: *Contaminated Land: Reclamation and Treatment.* Smith, M.A. (ed.) Plenum (London) 1985, pp 311–340.

19. BRITISH STANDARDS INSTITUTION. BS 1047: *Specification for air-cooled blastfurnace slag aggregate for use in construction.* BSI (London), 1983.

20. AMERICAN SOCIETY FOR TESTING AND MATERIALS. ASTM D 4792: *Standard test method for potential expansion of aggregates from hydration reactions.* ASTM (Philadelphia) – subject to annual review.

21. EMERY, J.J. A simple test procedure for evaluating the potential expansion of steel slag. In: *Proc. 1975 Annual Conference of the Roads and Transportation Association of Canada, Toronto 1975.*

Appendix 8 Important concepts and terms for risk assessment

Pathway

Pathways are the routes by which hazards reach targets. Transport media are air, soil, surface water and groundwater. Figures A8.1 to A8.3 illustrate in more detail the pathways and factors to take into account when trying to establish the transport and fate of contaminants in the environment.

Human health risk

Human health risk is the probability of injury, disease or death resulting from exposure to a hazard under specified circumstances. Examples of the types of human health risks that may be relevant in the context of contaminated land are summarised in Box A8.1.

Box A8.1 *Examples of potential human health risks in contaminated land applications*

- Acute (short-term) risks to the workforce coming into close contact with hazardous substances. Consideration must be given to any attendant physical hazards (which themselves represent a major category of acute risks) since these may enhance the risk of harm from exposure to the substance; for example a cut may provide a ready pathway for substances to enter the body
- Chronic (long-term) risks to workers arising from contact with hazardous substances including, for example, potential carcinogens where damage may not materialise until some time after the exposure event
- Acute risks to members of the public, or other targets, in close proximity to the site due to large scale events, such as fire or explosion; the release or migration of potentially toxic gases and vapours; or direct contact with contaminants (e.g. during trespass)
- Chronic risks to nearby and remote populations, and other targets, due to the release of contaminants from the site over long time periods
- Acute risks to users of the site
- Chronic risks to users of the site.

Human health risks depend not only on the nature of the hazard (e.g. explosion, toxicity, corrosivity) but also the route by which the exposure takes place and the duration of that exposure. Three main parameters are important:

- the **exposure** of the individual to the hazard
- the '**dose**' received
- the **level of toxicity** or **effect** associated with the hazard.

All human activities carry some degree of risk. Some of these risks are known with relative certainty because sufficient data have been collected to establish their occurrence (see

Tables A8.1 and A8.2 for the risk attached to some common human activities). However, human health risks associated with exposure to chemical and other contaminants in the context of their presence on a contaminated site are not known with certainty in many cases. This reflects a lack of basic toxicological data relating damage effects to dose (**dose-response relationships**), and/or because typical or anticipated routes of exposure in the context of contaminated ground have not been fully quantified.

Table A8.1 *Risk of death from a range of common causes*

Causes	Risk of death per million persons per year	Derived lifetime risks**
All causes (mainly illness from natural causes)	11 900	8.3×10^{-1}
Cancer	2 800	2.0×10^{-1}
All violent causes (accidents, homicide etc)	396	2.8×10^{-2}
Road accidents	100	7.0×10^{-3}
Accidents in private homes (for occupants)*	93	6.5×10^{-3}
Fire or flame (all types)*	15	1.0×10^{-3}
Drowning*	6	4.2×10^{-4}
Gas incident (fire, explosion, poisoning)	1.8	1.3×10^{-4}
Excessive cold*	8	5.6×10^{-4}
Lightening	0.1	7.0×10^{-6}
Accidents at work — risks to employees		
Deep-sea fishing (UK vessels)	880	6.0×10^{-2}
Coal extraction and manufacture of solid fuels	106	7.4×10^{-3}
Construction	92	6.4×10^{-3}
All manufacturing industry	23	1.6×10^{-3}
Offices, shops, warehouses etc	4.5	3.1×10^{-4}
Leisure — risks to participants during active years		
Rock climbing (assumes 200 hrs/yr)	8000	$1.6 \times 10^{-1}(20\ y)$
Canoeing (assumes 200 hrs/yr)	2000	$0.4 \times 10^{-1}(20\ y)$
Hang-gliding (average participant)	1500	$0.3 \times 10^{-1}(20\ y)$

Notes:

Risk figures are given as the average of the whole population of Great Britain, except where there is a specific small group exposed (e.g. rock climbing). The figures are given as the chance per million that a person will die from that cause in any one year, averaged over a whole lifetime (except where otherwise stated).

After HSE publication, 'Risk criteria for land-use planning in the vicinity of major industrial hazards' (1989). Sources quoted in original are HSE document on 'Tolerability of risk from nuclear power stations,' except * from OPCS Monitor series DH4 No 11, 1985, and Registrar General for Scotland, Annual Report, 1985.

** derived by authors for 70 year life (except where indicated) for this report assuming risk is not age related

Although considerable data have been gathered on the risks of some types of exposures (e.g. annual risk of death from intentional overdoses, or accidental exposure to drugs, pesticides and industrial chemicals), such data are generally restricted to acute poisoning incidents. In these situations, a single high exposure to the substance is likely to result in an immediately observable form of injury, leaving little doubt as to the cause. In the context of contaminated sites, exposure to hazardous substances is likely to be more complex with site users and neighbours typically exposed to extended or continuous sub-acute concentrations of hazardous substances. Quantification of human health risks is discussed further in Appendix 9

In most practical assessments of the risks to human health arising from contaminated sites, risks are expressed in terms of **'individual risk'** i.e. the risk that an identified adverse effect will occur in relation to an individual exposed to the hazard; for example that exposure will lead to an incremental lifetime increase in cancer of 1×10^{-4} or that the probable daily intake of a toxic substance is 1.5 times the Acceptable Daily Intake (ADI). For any particular hazard, it is usual to consider the 'typical' member of the most exposed group. For a residential development for example, the most vulnerable individuals at risk due to ingestion of soil and dust are young children who spend a high proportion of their time at home.

Ecological risks

A wide variety of ecological risks (eco-risks) may be posed by the presence of contamination on a site. Risks may relate to:

- populations (a group of organisms of the same species)
- communities (the combination of populations living in an ecosystem)
- ecosystems (communities and the physical environment in which they live).

Table A8.2 *Risks associated with travel*

Numerical estimates of risk of death from various forms of travel:

Number of deaths of passengers per 1000 million kilometres travelled	
Rail travel	0.45
Public road services	1.20
Scheduled air services	1.40
Car or taxi	7.00
Motor cycles	359.00

The estimation of ecological risks, even to a single species, is a complex process. For example in the case of a predator species, transfer factors between species, the behaviour of contaminants in different species (e.g. concentration in different organs) and feeding ranges may all have to be taken into account. Usually an attempt is made to identify the most sensitive target(s) in order to control the amount of investigation required (see Appendix 12).

Safety

The term 'safe' is sometimes taken to mean 'without risk' but is better regarded as a state in which deliberate efforts are made to minimise risks. It is not possible to define conditions under which exposure to a substance is absolutely without risk of any kind although it is possible to state that the risks are so low as to be of no practical significance to the population concerned. Such an approach has been taken in respect of the safety of chemicals in foodstuffs, drinking water etc., where a safe level is defined as that condition of exposure under which there is a 'practical certainty' that exposed individuals will not suffer harm as a result of exposure. Such levels usually incorporate large safety factors such that even where specified exposure levels are exceeded, the risk of harm remains relatively low.

Toxicity

Substances are often classified according to whether they are considered 'toxic' or 'non-toxic'. In the context of risk assessment, this approach is of relatively limited value since all substances can be made to produce a toxic response under some conditions of exposure. In terms of risk assessment, therefore, the important issue is not simply the degree of toxicity exerted by the substance but also the risk that the toxic properties will be realised under certain defined conditions of exposure.

In many cases the toxicity of a substance relates to both the chemical form of a substance and the exposure pathway; for example Cr^{VI} is a human carcinogen when inhaled (air pathway) but not when ingested (oral pathway).

Information on the toxic, and other properties, of substances has been collected in a number of different ways, e.g. by recording the outcome of accidental exposures, animal experimentation, epidemiological studies of exposed individuals, and experimental work involving surrogate materials such as cell cultures, micro-organisms etc.

Animal studies usually commence with a consideration of acute effects and progress towards the examination of chronic effects. A frequently used measure of the toxicity of a substance is the LD_{50}, which is the dose required to produce death in 50% of the exposed individuals. While such measures provide a useful indicator of the relative toxicity of substances, it is important to remember that there are limits to which the results of animal or 'indicator species' experiments can be usefully extrapolated to the human exposure situation.

Exposure and dose

The amount of a substance present in the contaminated medium is measured as a concentration, the '**exposure concentration**', while the amount of the substance reaching the target is termed the '**dose**'. For human or other animals this is measured as the amount of the substance per unit of body weight per unit of time (see Appendix 9). Analogous concepts can be applied to other potential targets.

Exposure assessment

In an exposure assessment the amount of exposure of the target is determined. This addresses how often, for how long, and over what period the exposure has occurred and is likely to continue, or is likely to occur in the future. It requires consideration of all routes of exposure as shown in Figures A8.1 to A8.3.

Dose-response relationships

A dose-response relationship defines the effect that a given dose (or equivalent quantity for non-animal targets) has on the target. The aim will usually be to identify a dose that produces 'no observable adverse effect' or in broader terms 'no unacceptable adverse effect' (for example the effect may be transient and aesthetic, such as high noise levels or release of unpleasant odours for a short period of time, rather than lasting – see also the discussion in Appendix 12 on the effects of contaminants on ecosystems). For a discussion of dose-response relationships in humans see Appendix 9.

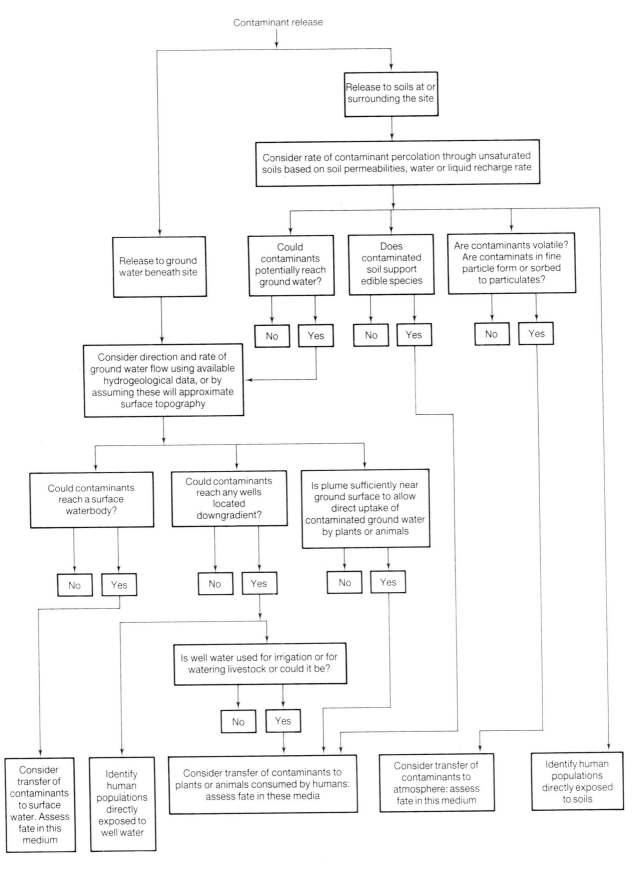

Figure A8.1 *Flow chart for fate and transport assessments: surface water and sediment*

Toxicity assessment for human health risks

In toxicity assessments the adverse effects of contaminants (acute/chronic, non-carcinogenic/carcinogenic/mutagenic etc.) are first identified (see also Appendix 9). Published standards and guidelines, and other sources, are then employed to identify dose-response relationships and acceptable intakes against which the estimated doses received by targets are compared.

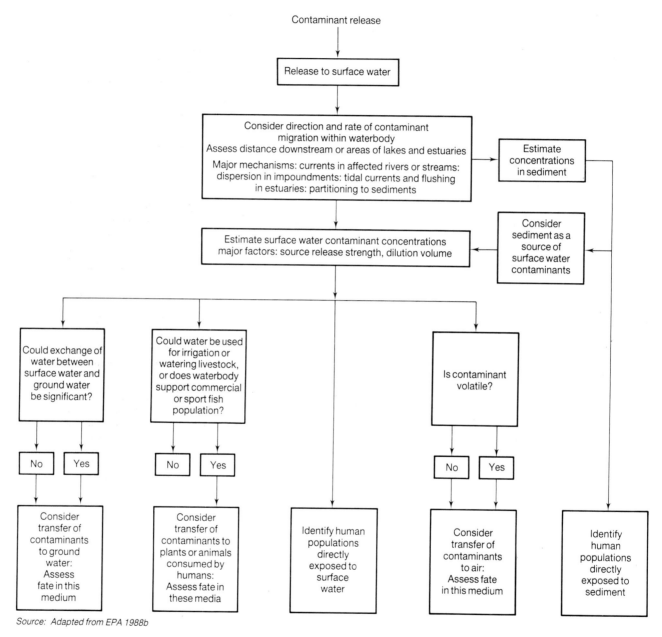

Source: Adapted from EPA 1988b

Figure A8.2 *Flow chart for fate and transport assessments: soils and groundwaters*

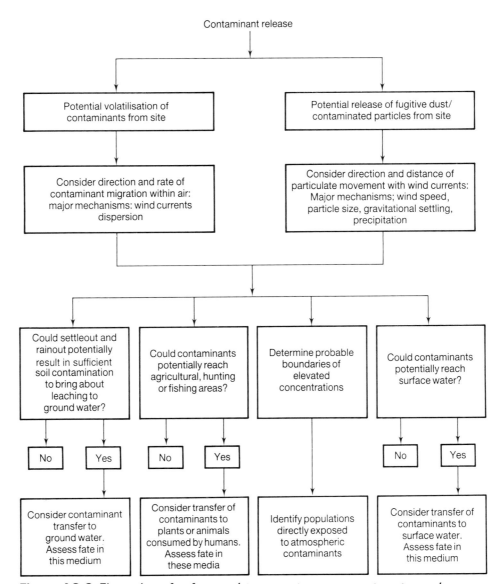

Figure A8.3 *Flow chart for fate and transport assessments: atmosphere*

CIRIA Special Publication 103

Appendix 9 Quantifying human health risks

A9.1 EXPOSURE AND DOSE IN HUMANS

A9.1.1 Exposure and dose

Certain standard assumptions are used to estimate dose. For example, it may be assumed that a typical adult male body weight is 70 kg, that of a female adult is 50 kg and that of an infant 10 kg. Other assumptions, or measured factors, are used to determine the quantity of the contaminated substance likely to be taken into the body over a defined period of time (see references 1 to 3 for formulae and standard assumptions used for a variety of exposure routes).

This relatively simple relationship becomes more complicated as additional factors are taken into account. For example, the individual may be exposed to the contaminant via a number of routes. Dose estimates, therefore, may need to take into account contributions of the substance from a variety of sources. For a child, for example, possible exposure to a contaminant via ingestion of contaminated foodstuffs may need to be supplemented by contributions from other ingestion routes, such as the transfer of contaminants from the ground via dirty fingers or on food items.

The question of whether simple summation of the dose from all possible sources accurately reflects the true impact of the substance on the individual must be addressed. Mode of entry may be significant in this respect since substances differ in their toxic and other effects depending on the route of access to the target. For example, chromium is carcinogenic when inhaled but appears not to exert a similar effect when ingested. Provided the effect of the substance is systemic (i.e. affects the whole body), rather than confined to the point of exposure, the summation of dose is usually acceptable.

Not all the substance taken in by the individual is likely to be absorbed and is therefore capable of exerting an effect at the target organ level. Some of the material may be excreted, immobilised or detoxified and it important therefore to distinguish between total dose and absorbed dose. In practice, data on absorption characteristics of substances are limited and it is rarely possible to allow for partial absorption in dose estimates.

A9.1.2 Dose-response relationships

A variety of complex dose-relationships exist. For substances which do not display carcinogenic properties, or for the non-carcinogenic properties of carcinogens, dose-response evaluation consists of describing observed dose-response relationships and identifying experimental NOAELs (No Observable Adverse Health Effect Limits). NOAELs can be used to derive RfDs (Reference Doses). These are derived by dividing the NOAELs by an appropriate safety or uncertainty factor. For a detailed discussion of dose-response relationships reference should be made to standard texts (e.g. reference 4).

A9.2 QUANTIFYING HUMAN HEALTH RISKS

A9.2.1 Risks for individual substances

Risks from carcinogens are estimated as the incremental probability of an individual developing cancer over a lifetime as a result of exposure to a potential carcinogen (i.e. incremental or excess individual lifetime cancer risk). This calculation depends on a knowledge of the relationship between exposure and effect for individual compounds. It is usual to assume that there is no safe exposure level, i.e. a level at which there is no adverse effect. Detailed consideration of such calculations is beyond the scope of this Report.

Risks of non-carcinogenic effects are usually expressed by comparing the exposure (E) level over a set period of time (e.g. lifetime) with a reference dose (RfD) for a similar period of time. This ratio of exposure to toxicity (E/RfD) is usually[1] called a hazard quotient. A low value implies a low risk.

A9.2.2 Aggregate effects of multiple substances

A range of compounds will be present on most sites presenting a range of hazards. The effects of exposure to such compounds may be additive or synergistic i.e. two or more substances combine to produce more than proportionate effects. In some cases, the presence of one substance may provide the trigger for action by another which is otherwise harmless. Thus estimating risk or hazard by considering one substance at a time might significantly underestimate the risks associated with simultaneous exposure to several substances.

The US Environmental Protection Agency has published[1] standardised methods for indexing such additive effects. Hazard quotients for individual substances are combined into hazard indices. Non-cancer risks are usually characterised in terms of chronic exposure (7 years to life time), sub-chronic exposures (two weeks to seven years) and shorter term exposures (less than two weeks).

A decision on whether to combine hazard quotients may take into account whether the same or different organs are at risk.

A9.2.3 Combining risks across exposure pathways

Combination of exposures from more than one pathway requires identification of pathways that are likely to lead to additive exposures and decisions about the probability of individuals or small target populations being exposed by each route. The US EPA has made standardised procedures available for this purpose[1].

A9.2.4 Acceptable risks

A judgement must be made about the risks which are 'acceptable' in any particular circumstances. Acceptable risks to the general public (which includes children/pregnant women, the infirm etc.) will differ from those for a healthy member of the workforce.

It is not appropriate here to define acceptable risks. However, incremental individual lifetime cancer risks in excess of 10^{-4} are unlikely to be acceptable and 10^{-6} to 10^{-7} to vulnerable receptors is likely to be considered a desirable target in many circumstances. For non-cancer hazards a combined hazard index in excess of 1.0 will often be regarded as unacceptable (as noted above separate indices for different organs may sometimes be appropriate and aid in the risk evaluation). However, the significance of such estimated risks will depend on the size of the population exposed, currently or in the future, and the uncertainties of the risk assessment. For a useful discussion of the acceptability of different types of risk see references 5 and 6.

REFERENCES

1. UNITED STATES ENVIRONMENTAL PROTECTION AGENCY. *Risk assessment guidance for Superfund, Volume I: Human health evaluation manual (Part A)*. US Office of Emergency and Remedial Response (Washington DC), EPA/540/189/002, 1989

2. Guidelines for exposure assessment. *Notice Federal Register*, Vol 57, No 104 (29 May 1992), pp 22888-22938

3. STATE OF CALIFORNIA DEPARTMENT OF HEALTH SERVICES. *Technical standard for determination of soil remediation levels*. SCDHS, 1990

4. UNITED STATES ENVIRONMENTAL PROTECTION AGENCY. *The risk assessment guidelines of 1984*. US Office of Health and Environmental Assessment (Washington DC), EPA/600/8−87/045, 1987

5. HEALTH AND SAFETY EXECUTIVE. *Risk criteria for land-use planning in the vicinity of major industrial hazards*. HMSO (London), 1989

6. HEALTH AND SAFETY EXECUTIVE. *The tolerability of risk from nuclear power stations*. HMSO (London), 1988

Appendix 10 Examples of modelling in risk assessment

A10.1 APPLICATION OF GROUNDWATER MODELLING AND RISK ASSESSMENT AT A PETROL STATION

When conducting human health risk assessments using groundwater/contaminant transport models it is important to ensure that they produce realistic results. Simple models, which take account of advection (movement with the flow of groundwater) and dispersion only, generate higher-than-actual estimates of contaminant concentrations down gradient of the source (see Volume VI). This can lead to overly conservative risk estimates and consequently, to an overly conservative remediation strategy. Figure A10.1 shows the different results obtained using simple and complex groundwater models for a typical petrol constituent at a point in time.

The example presented in Box A10.1 shows how groundwater modelling techniques which take the interaction of contaminants with soil (e.g. adsorption) into account can be used as part of a quantified risk assessment to demonstrate the absence of a significant health risk.

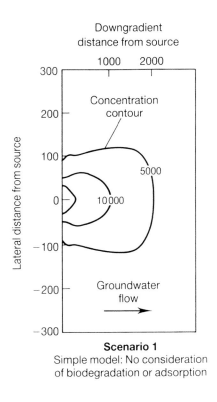
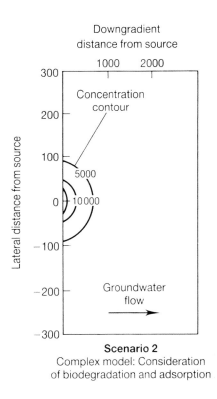

Figure A10.1 *Effects of adsorption and biodegradation on groundwater modelling results*

> **Box A10.1** *Application of groundwater modelling at a petrol station*
>
> An investigation of contamination arising from leaking underground storage tanks at a petrol service station revealed the presence of benzene, toluene, ethylbenzene, xylenes, methyl terbutyl ether and other petroleum hydrocarbons, in solution in the groundwater. A risk assessment was carried out to quantify and evaluate the contaminant levels that might potentially reach the people living closest to the site. The risk assessment was conducted for both current and future exposure pathways. A groundwater/contaminant transport model, capable of accounting for adsorption and biodegradation, was used to predict future groundwater contaminant concentrations at points distant from the source. The predicted concentrations were then converted into chronic daily intakes and evaluated in accordance with US Environmental Protection Agency protocols.
>
> The results of the risk assessment indicated that the groundwater contamination did not pose an unacceptable risk to people at the time of the study or in the future. Consequently, the relevant regulatory bodies accepted that no active remediation was required.
>
> The cost of the quantitative risk assessment was approximately US $15 000. The cost of active remediation was estimated to be US $200 000 to $400 000.

A10.2 MODELLING AIR EMISSIONS AS PART OF A RISK ASSESSMENT

Consideration of air quality impacts may be required :

- as part of a planning application (DOE circular 15/88, EEC Directive 85/337/EEC)

- to otherwise satisfy the local Environmental Health Officer about possible adverse health effects or odour nuisance on the general public

- by the Health and Safety Executive in relation to potential impacts on workers directly engaged in the planned operation, on workers in adjacent premises and on the general public.

Modelling is an essential tool in the estimation of such impacts but needs to be applied in conjunction with field monitoring to establish base line conditions, to supplement modelled predictions and to provide a measure against which the accuracy of the predictions can be assessed.

Because of spatial and temporal limitations, monitoring data are not normally sufficient on their own to demonstrate potential exposure characteristics; models can be used to interpolate values at unmonitored locations and to predict the consequences of changing weather conditions etc.

A range of models has been developed to deal with different situations. No single model can cope with all eventualities. In the case of odour nuisance and short term toxicity effects, the model must be capable of taking short term variations (caused for example by local atmospheric effects) into account. Simulation dispersion modelling based on the Gaussian plume is powerful technique in the assessment of potential incremental impacts of emissions on ambient concentrations of pollutants because it uses actual variability in meteorological data to predict dispersion patterns.

A major limitation of such models is that they require long periods of historic or on-line meteorological data, preferably taken in close proximity to the point of emission or at least in similar nearby terrain. The model is inevitably a discrete rather than a continuous simulation and relies on a simplistic view of meteorological conditions. Problems are experienced in modelling calm wind conditions. Atmospheric reaction chemistry is not normally taken into account.

A variety of modelling packages are available. The most widely accepted and used are the UNAMAP (Users Network for Applied Modelling of Air Pollution) series produced by the US Environmental Protection Agency. The series includes both screening and refined models. Screening models tend to produce conservative predictions. They can be run using fairly limited data and with short computing times to determine the potential for a problem which can then be explored further with a more refined model.

In order to be able to run a model, data are required on area, site and source characteristics. Area characteristics include such aspects as topography, local meteorology (with data frequently available in digital form for direct application), location and heights etc. of neighbouring buildings, and other potential emission sources. Critical site characteristics include the proximity of buildings to the point of emission since nearby buildings can markedly affect near ground wind speeds and directions. Source characteristics required include location of source (height etc.), size of opening etc; height above sea level; physical characteristics of the emission itself such as temperature, velocity and chemical composition, and hence the rate of emission of individual substances.

Output data are presented as time averaged concentrations at specific receptor points. The results are usually presented as plots of equal concentration (isopleths) overlaid on a map of the area.

Concentrations data must then be compared with suitable criteria. In the absence of specific UK or EEC air quality standards for organic compounds, assessment criteria are usually based on one or more of the following:

- odour threshold concentrations for perception and/or recognition

- health risk threshold, based on rule of thumb safety factors, e.g. one fortieth or a lesser fraction of the Occupational Exposure Standard (see Section 6)

- World Health Organisation Air Quality Guideline values which include health and odour based values

- typical ambient concentrations reported in the literature.

Short term models using hourly meterological data can predict maximum ground level concentrations at any specific receptor up to 50 km from the source at averaging times of between one hour and one year. The frequency with which these particular criteria are exceeded can then be predicted.

In general long term modelling predictions are better than short term types and predictions of the locations of maximum concentrations are more accurate. Errors in predicting maximum concentrations within an area can be as much as ± 10 to 40%, and $\pm 50\%$ in terms of maximum concentrations on an hourly basis at a given point. The predictions of near source concentrations tend to be less reliable than those for greater distances.

Special models are required for explosive events/sudden excursions.

Validation of the model outputs using a well designed and executed measurement programme is essential.

Appendix 11 International guidelines and standards

A11.1 SOILS, FILLS AND SEDIMENTS

A number of countries and jurisdictions have promulgated dedicated guidelines or criteria for contaminants in soils[1]. Where ICRCL guidelines for a contaminant are not available, international guidelines may be helpful in deriving site-specific action and remediation values. Note that it is essential to understand the technical basis of the values and the policy and legal context for application in the country of origin. Tables of values should not be used without reference to accompanying written guidance on their application.

Some international values have been set on a judgemental basis, others through application of more or less complex risk assessment models. Many are subject to periodic review and amendment and users should ensure that they use only the most recent editions. The Dutch and Canadian values can be regarded as being fairly typical of other international guidelines.

The original Dutch ABC values (see Box A11.1) were introduced in support of policies for soil protection and the preservation of the multi-functionality of soils[2]. These original Dutch values have been employed directly in some other countries and have provided the base set of figures for generic values adopted in a number of other jurisdictions with appropriate amendment to suit local conditions. However, most jurisdictions which have adopted, or adapted, the Dutch ABC list do not apply the same multi-functionality policy.

> **Box A11.1** *The Dutch ABC Values*
>
> The Dutch 'ABC' values were introduced in 1983 as mandatory standards in support of the Soil Protection Guideline[6] and were to be used in conjunction with other criteria within the guideline. They were not regarded as generic soil quality criteria.
>
> A-values were 'target values' (remediation standards) and, for naturally occurring contaminants, were originally set at background levels typical of 'uncontaminated' rural soils in the Netherlands. For most of the more unusual organic species listed, A-values tended to be set at, or close to, analytical detection levels. A-values were replaced in 1987 by Reference Values for a 'good quality soil.' These take soil properties (e.g. clay and organic matter content) into account.[3]
>
> The B-values represented that concentration above which additional investigation of the site is required.
>
> C-values represented concentrations above which remedial action was considered essential (intervention value). If remedial action was triggered as a result, the aim was to achieve levels of contaminants in treated soil consistent with A-values (now Reference Values).
>
> In terms of the actions prompted by the guidelines, the Dutch B-value corresponded to the ICRCL threshold trigger value and the C-value to the action trigger value, the main difference being that the ABC values were mandatory standards whereas the ICRCL values are guidelines only.

A11.1.1 Dutch values

The revised Dutch system replaces the ABC values with just two sets of numbers :

- a Target Value (A) for a 'good quality' soil which takes soil characteristics (e.g. organic matter content, clay content) into account
- an Intervention Value (C) which will take both human health and ecological risks into account.

The definition of a good quality soil is one that:

- *poses no harm to any use by human beings, plants and animals*
- *can function without restriction in natural cycles*
- *does not contaminate other parts of the environment.*

For a description of the derivation of the new A- and C-values see references 2 to 5.

Box A11.2 *Application of Dutch Soil Protection Standards*

General principles

To assess the extent and seriousness of soil pollution a stepwise investigation is carried out. After each stage a decision is taken as to whether to proceed to the next stage. The decision is based on qualitative criteria given in the Soil Protection Guideline[6] and the trigger values (A and C) for the concentrations of contaminants in soil and groundwater. The stepwise process involves:

Preliminary investigation: Desk study followed by limited sampling programme. If concentrations exceed [(A+C)/2], further investigations are necessary. If concentrations are close to the A-values, the site is considered clean.

Further investigation: Further sampling etc. to see if intervention values (C-values) are exceeded and to determine spatial distribution. If intervention values are exceeded remediation is considered necessary and the site is 'registered.' If there is no exposure and the contaminants are not mobile, the situation is not considered urgent and no action is required unless there is a change in land use or the contaminants are mobilised.

Clean-up investigation: This investigation explores possible remedial options, taking other technical and environmental aspects of available techniques into account. Dutch government policy is to deal with contaminants in an absolute sense unless costs or technical problems make such an option impossible in practice. Should this be the case, appropriate containment, isolation, partial remediation and control measures (e.g. restriction of land use) may be employed. If clean-up is to be carried out then the 'target' is to achieve the A-value.

The values are to be used within a prescribed framework of site investigation and assessment. This defines the level of investigation and other responses to be made to observed levels of contamination. The B-value of the old system (see Box A11.1) is to be replaced by a new value [(A+C)/2] that triggers detailed investigation (see Box A11.2). Note that while the standards indicate a potential need for action, the urgency of the action will be influenced by other factors, in particular current exposure, and anticipated future exposure associated with any planned change in land use.

The values and associated decision-making are intended to be used within a general policy aiming at the conservation and restoration of the soil (multi-functionality). In practice, however, a flexible interpretation of the values may be permissable at site level provided agreement with the regulatory authorities can be reached (see Box A11.2). General guidelines describing environmental, technical and financial conditions for a non-multifunctional approach are under development.

A11.1.2 Canadian criteria

The Canadian Federal Government has recently introduced[7,8] a range of 'environmental quality criteria for contaminated sites' to provide:

> *'general technical and scientific guidance to provincial, federal, territorial and non-governmental agencies in the assessment and remediation of contaminated sites. They serve as benchmarks against which to assess the degree of contamination at a site and to provide guidance on the need for remediation, the establishment of remediation goals and strategies, and verification of the adequacy of remedial actions. Most important, they constitute a common scientific basis for the establishment of remediation objectives for specific sites. Variations in local conditions, existing guidelines and standards, and technological, socioeconomic, or legal considerations may affect they way they are applied at the site-specific level. A detailed consideration of these site-specific factors will therefore usually be required before regulatory requirements or remedial actions are finalized.'*

Three terms are defined:

1. **Criterion**: a generic numerical limit or narrative statement intended as general guidance for the protection, maintenance, and improvement of specific uses of soils and water.

2. **Objective**: a numerical limit or narrative statement that has been established to protect and maintain a specified use of soil or water at a particular site by taking into account site-specific factors.

3. **Standard**: a legally enforceable numerical limit or narrative statement, such as in a regulation, statute, contract or other legally binding document, which has been adopted from a criterion or an objective.

Two types of criteria are provided:

- Assessment criteria which serve as benchmarks against which to assess the degree of contamination at a site and to determine the need for further action. The interim assessment criteria are approximate background levels or analytical detection limits and are intended to provide general guidance only.

- Remediation criteria to be used as benchmarks to evaluate the need for further investigation or remediation with respect to a specified land use. They are intended for generic use and do not address site-specific factors. They are considered generally protective of human and environmental health for specified uses of soil and water at contaminated sites, based on experience and professional judgement.

  Remediation criteria for soil are provided for agricultural, residential/parkland, and industrial/commercial land uses. Remediation criteria for water, taken from existing Canadian guidelines[9,10] are provided for specified uses of water (freshwater aquatic life, irrigation, livestock watering, drinking water) likely to be of concern at contaminated sites.

The Canadian guidelines recognise that the generic criteria cannot be adopted directly as measures of remediation performance, and that due consideration must be given to site-specific factors. Therefore a two-tiered approach has been adopted using generic criteria as a reference point and then setting site specific remediation objectives. Depending on the circumstances these may:

- involve adoption or adaption of existing criteria (*criteria-based approach*)
- be based on ecological/human health risk assessment (*risk-based approach*).

The framework is thus intended to take advantage of the consistency, economy and speed of the generic approach, while maintaining some flexibility in setting site-specific objectives and the option of risk assessment when warranted.

The interim criteria have been drawn up following a review of guidance available in other jurisdictions and of other relevant literature. They are supported by a programme to progressively devise criteria that will take both human and ecological risks into account[7,11].

A11.1.3 Australian and New Zealand guidelines

The Australian and New Zealand Environment and Conservation Council and the (Australian) National Health and Medical Research Council have provided guidance[12] for the assessment and management of contaminated sites. It covers site identification, investigation and assessment and provides for the development of 'public health based guidelines' and 'environment based guidelines.' An 'investigation level' is defined as providing 'a trigger to assist in judging whether a detailed investigation of a site is necessary.' Separate health and environmental investigation levels are provided. Selection of the most appropriate is decided on a site-specific basis.

Environmental soil quality guidelines for background levels (A) and environmental investigation (B) values (also termed investigation threshold levels) have been provided for a range of inorganic and organic contaminants. The list of public health investigation levels in the guidance document is more restricted but is to be extended.

A11.2 SURFACE AND GROUNDWATERS

A11.2.1 Dutch values

Under the revised system A- and C-values (intervention values) comparable to those for soils are provided for waters. The A-values have been derived from existing standards for drinking and surface water; C-values from consideration of human and ecological risks.

A11.2.2 Canadian national criteria

The basis of the interim Canadian environmental quality criteria for contaminated sites has been described above. Assessment criteria are provided for water. Remediation criteria are provided for four water uses: freshwater aquatic life, irrigation, livestock watering and drinking water. The interim criteria have been derived from existing Canadian guidance on water quality[9,10]. Proposals for protocols for the derivation of guidelines for agricultural water uses and for protection of aquatic life have been prepared[13,14].

A11.2.3 British Columbia

British Columbia has produced two standards of relevance to water quality[15]. The first relates to waters intended for drinking purposes and the second to the quality of the aqueous environment:

1. The B_{DW} specifies the concentration of a substance above which remedial works are required where water is intended for human consumption.

2. The B_{DS} is the criterion for water-based discharges required to protect aquatic life. For discharges containing constituents at concentrations less than level B_{DS} remedial works are not required provided the receiving water represents only a habitat for aquatic life. Contaminant concentrations exceeding B_{DS} require further investigation to assess the impact of the substances on water quality and to determine appropriate action.

REFERENCES

1. *Proceedings of a Conference on Developing Clean-up Standards for Contaminated Soil, Sediment and Groundwater: How Clean is Clean?* Water and Environment Federation (Alexandria, Virginia), 1993

2. VEGTER, J.J. Developments of soil and groundwater clean-up standards in the Netherlands. In: *Proceedings of a Conference on Developing Clean-up Standards for Contaminated Soil, Sediment and Groundwater: How Clean is Clean?* Water and Environment Federation (Alexandria, Virginia), 1993, pp 81-92

3. MOEN, J.E.T. Soil protection in the Netherlands. In: *Proceedings of the Second International TNO/BMFT Conference on Contaminated Soil.* Kluwer (Dordrecht), 1988, pp 1495-1504

4. DENNEMAN, C.A.C. and ROBBERSE, J.G. Ecotoxicological risk assessment as a base for development of soil quality criteria. In: *Proceedings of the Third International KfK/TNO Conference on Contaminated Soil.* Kluwer (Dordrecht), 1990, pp 157-164

5. MINISTRY OF HOUSING, PHYSICAL PLANNING AND ENVIRONMENT. *Environmental quality standards for soil and water.* MHPPE/VROM (The Hague), 1991

6. MINISTERIE van VOLKSHUISVESTING RUIMTELIJKE ORDENING EN MILEUBEHEER. *Soil protection guideline.* Staatsuitgeverij (s-Gravenhage), 1990

7. CANADIAN COUNCIL OF ENVIRONMENT MINISTERS. *Interim CCME environmental quality criteria for contaminated sites.* Report No. CCME EPC-CS34. CCEM (Winnipeg), 1991

8. HOFMAN, E.L. *et al.* Setting goals for contaminated sites: towards a nationally consistent approach in Canada. In: *Proceedings of a Conference on Developing Clean-up Standards for Contaminated Soil, Sediment and Groundwater: How Clean is Clean?* Water and Environment Federation (Alexandria, Virginia), 1993, pp 69-91

9. CANADIAN COUNCIL OF RESOURCE AND ENVIRONMENT MINISTERS. *Canadian Water Quality Guidelines.* CCREM (Winnipeg), 1987

10. HEALTH AND WELFARE CANADA. *Guidelines for Canadian Drinking Water Quality.* 4th Edition. Canadian Government Publishing Centre (Ottawa), 1989

11. CANADIAN COUNCIL OF ENVIRONMENT MINISTERS. *A protocol for the derivation of ecological effects-based and human health based quality criteria for contaminated sites.* CCME (Winnipeg), (in preparation, 1993)

12. AUSTRALIAN AND NEW ZEALAND ENVIRONMENT AND CONSERVATION COUNCIL/ NATIONAL HEALTH AND MEDICAL RESEARCH COUNCIL. Australian and New Zealand guidelines for the assessment and management of contaminated sites ANZECC/ NHMRC, 1992

13. CANADIAN COUNCIL OF ENVIRONMENT MINISTERS. *Proposed protocols for derivation of water quality guidelines for the protection of aquatic life.* Water Quality Branch, Environment Canada (Ottawa), 1991

14. CANADIAN COUNCIL OF ENVIRONMENT MINISTERS. *Proposed protocols for derivation of water quality guidelines for the protection of agricultural water uses.* Water Quality Branch, Environment Canada (Ottawa), 1991

15. MINISTRY OF THE ENVIRONMENT. *Criteria for managing contaminated sites in British Columbia.* Ministry of the Environment (Victoria, BC), 1989

Further reading

US ENVIRONMENTAL PROTECTION AGENCY. *Protection EPA – State Soil Standards Conference.* USEPA (Washington DC), 1922. EPA 1540/R-92/005.

MINISTRY OF HOUSING, SPATIAL PLANNING AND THE ENVIRONMENT. *Environmental quality objectives in the Netherlands.* (The Hague), 1994

MINISTRY OF HOUSING, SPATIAL PLANNING AND THE ENVIRONMENT. *Soil Protection Act 1994.* (The Hague), 1994

Appendix 12 Ecological risk assessment

A12.1 INTRODUCTION

A12.1.1 Objectives of ecological risk assessment

Ecological risk assessment (some-times also called eco-risk assessment or environmental evaluation) is a qualitative and/or quantitative appraisal of the actual or potential effects of a contaminated site on plants and animals, other than humans and domesticated species.

It provides decision makers with information on risks to the natural environment posed by contaminants or action intended to remediate or develop sites. It helps to reduce the inevitable uncertainty about possible environmental impacts and puts bounds on those uncertainties. Some uncertainties will always remain: ecological risk assessments are not research projects and they cannot answer long-term research needs.

Not all sites will require eco-risk assessments. Many are in industrial areas where there is little wildlife or are remote from any body of surface water which might be affected by migrating contaminants. However, some contaminated sites may have an important ecological value: some have been designated as Sites of Special Scientific Interest (SSSIs) because of the rare plant communities that have become established because of, or despite, hostile or unusual conditions.

The ecological aspects of importance are:

- the protection of ecological targets at or near the site
- the effects of contaminants on these targets
- the effect of remedial action.

Ecological assessments may be carried out before, during and after remedial action or development.

During the course of remediation/development decisions have to be made about:

- whether remedial action is required on ecological grounds
- the scope of any ecological assessment
- monitoring strategies to determine the progress, effectiveness and impacts of any remedial/development works
- the likely ecological impact of any remedial/development works.

As with all other forms of site investigation, any detailed assessment should be preceded by a preliminary investigation in which available information is collected and initial field observations made. These data can be used to decide whether a detailed assessment is required, and if so, the scope of the work and how it should be conducted.

Ecological and human health assessments are parallel activities in the evaluation of contaminated sites. As Figure A12.1 illustrates, most of the data and analyses relating to the nature, fate and transport of contaminants can be used for both assessments, although the importance of different pathways will vary. Data on bio-accumulation in wild or game species can be helpful in human health assessments and vice-versa. The health of people and domesticated animals is inextricably linked with the quality of the environment shared with other species. Information from ecological studies may point to new or unexpected exposure pathways for human populations, and health assessments may help to identify environmental risks.

Ecological assessment may also be useful in identifying strategies for monitoring the progress and effectiveness of remediation activities at or near the site. For example, toxicity tests on soils, sediments, and water may be used to supplement chemical testing in establishing remediation criteria. Toxicity tests may be more sensitive to low levels of contaminants than other monitoring methods indicating the toxicity of mixtures of contaminants more readily than other test procedures.

However, ecological risk assessment is more complex than human health risk assessment (there are potentially more targets, pathways and responses involved) and there is less guidance, and fewer consensus models and standard assumptions available than in either the human health or engineering risk assessment contexts.

A12.1.2 Sources of information

In the preparation of this Section, extensive use has been made of information produced by the US Environmental Protection Agency (EPA). This includes background information on ecological risk assessment for managers of site remediation projects[1]; a complementary volume on field and laboratory methods[2]; and a general 'framework for ecological risk assessment[3]'. This framework covers all situations in which such an assessment might be required (e.g. the release of chemicals to the environment, draining of wetlands, exploitation of 'renewable' resource such as a fishery) and both chemical and physical stressors. The framework is summarised briefly in Box A12.1; the concept of ecological risk assessment which forms the basis of the framework is described in Box A12.2. The framework and the application of ecological risk assessment to contaminated sites, including a number of case studies, are described in the proceedings of a workshop organised by the Water Environment Federation in 1993[4]. Further useful general discussion of eco-risk assessment in the context of contaminated sites is provided by Suter and Loar[5]. For a discussion of the special problems posed by sediments see Adams *et al*[6] and Burton and Scott[7]. Suter[8] provides a good general text on biological risk assessment (although not specific to contaminated sites); the practical aspects of assessing the ecological impacts of contaminated land are discussed by Maughan[9].

A12.1.3 Application of the guidance

In England and Wales, and depending on the scale and nature of the works and the after use of the site, an environmental (impact) assessment (EIA) may be required under the Town and Country Planning (Assessment of Environmental Effects) Regulations 1988 (similar requirements apply in Scotland and Northern Ireland). Ecological risk assessment as described here can make a valuable contribution to EIA; some of the guidance already available in the UK on EIA may also be helpful to the application of ecological risk assessment (see Volume XII).

Box A12.1 *The US Environmental Protection Agency's Framework for Ecological Risk Assessment*[3]

The framework is 'intended to offer a simple, flexible structure for conducting and evaluating ecological risk assessment within EPA.'

It comprises three major phases (see Figure A12.2):

- problem formulation
- analysis
- risk characterisation.

Problem formulation is a planning phase which establishes the goals, breadth and focus of the risk assessment. The output is a conceptual model identifying the environmental values to be protected (the *assessment endpoints*), the data needed, and the analyses to be used.

The analysis phase develops profiles of environmental exposure and the effects of the stressor. The exposure profile characterises the ecosystems in which the stressor may occur as well as the biota that may be exposed. It also describes the magnitude, and spacial and temporal patterns of exposure. The ecological effects profile summarises data on the effects of the stressor and relates them to the assessment endpoints.

Risk characterisation (embracing the processes of risk estimation and risk evaluation described in Section 5) integrates the exposure and effects profiles. Risks can be estimated using a variety of techniques including comparing individual exposure and effects values, comparing the distribution of exposure and effects, or using simulation models. Risk can be expressed in qualitative or quantitative terms, depending on the available data.

The framework also identifies several activities that are integral to, but separate from, the risk assessment as defined in the framework document. For example, discussions between the assessor and the risk manager, data acquisition and verification and monitoring studies.

Note that data acquisition is considered part of the risk assessment process in Section 5.

Box A12.2 *The concept of ecological risk assessment*[3]

Ecological risk assessment is defined as the process of evaluating the likelihood that adverse effects may occur or are occurring as a result of exposure to one or more stressors. A risk does not exist unless:

1. the stressor has the inherent ability to cause one or more adverse effects, and
2. it coexists with or contacts an ecological component (i.e. organism, population, community or ecosystem) long enough and at sufficient intensity to elicit the identified adverse effect.

Ecological risk assessment may evaluate one or many stressors and ecological components.

The results of the assessment may be expressed as probabilistic estimates of risk, or in deterministic or qualitative terms.

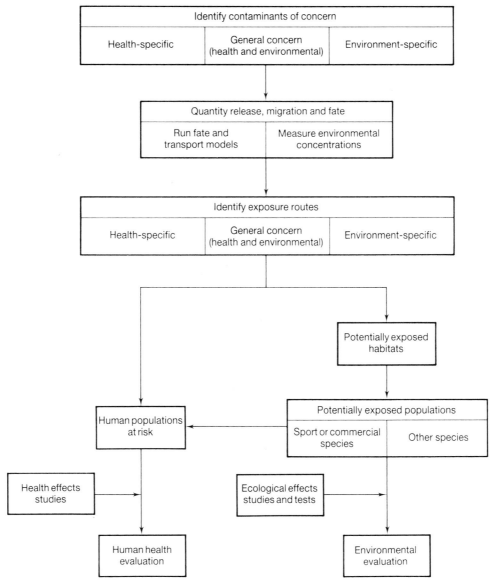

Figure A12.1 *Relationship between health and environmental evaluations/ecological assessments*

This Appendix does not provide a manual of how to carry out ecological risk assessments. It is intended to provide background information and a framework to inform and assist decision makers about the key issues that should be addressed. Specialist advice should always be sought. Specialist expertise in the following areas may be required for ecological risk assessment:

- environmental chemistry
- transport and fate modelling
- ecological toxicity
- ecology

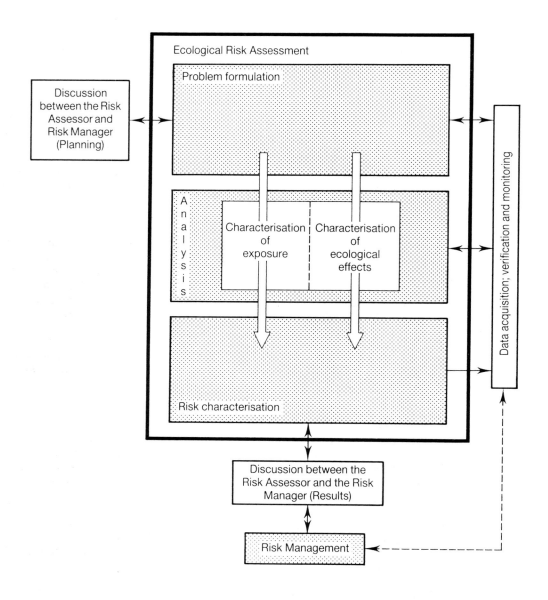

Figure A12.2 *Framework for ecological risk assessment*[3]

A12.2 BASIC CONCEPTS

Three levels of ecological organisation can be distinguished: populations, communities, and ecosystems. These terms are explained in more detail in Box A12.3.

> **Box A12.3** *Basic concepts of ecology*
>
> **Population**
>
> A population is a group of organisms of the same species occupying a contiguous area, and capable of interbreeding. Populations are normally described in terms of total numbers and density − the number of organisms per unit area/volume.
>
> Structural factors such as proportions of eggs, juveniles and adults are measured. Population growth and decline are determined by birth, death, migration rates etc.
>
> **Communities**
>
> All organisms and populations are members of communities. The interactions among populations and the physical and chemical features of the environment together determine a community's structure and geographical extent. The structure can be defined in a number of ways including the species present (the absolute and relative numbers of each); the food web or trophic structure (i.e. what species eat which other species, or who produces or consumes how much); the vertical structure of the vegetation etc.
>
> Most communities change seasonally or over longer cycles, evolving over long periods of time in a process known a succession e.g a meadow evolves into scrub and eventually into woodland. Species diversity is used to characterise and compare the structure and maturity of communities.
>
> **Ecosystems**
>
> Communities interact continuously with non-living components of the environment in an ecosystem: *'a functional system of complementary relationships and transfer and circulation of energy and matter.'* The ecosystem comprises all the living organisms, their remains, and the minerals, chemicals, water and atmosphere on which they depend for sustenance and shelter.
>
> Ecosystems are characterised in much the same way as communities in terms of: species composition and diversity, nutrient and energy flows, and rates of production, consumption, and decomposition. The ecosystem is considered to be the fundamental unit of ecology.
>
> Energy and matter flow through ecosystems in complex systems known as food chains (hierarchical assemblages) and food webs (interconnecting food chains).

A12.3 EFFECTS OF CONTAMINANTS ON ECOSYSTEMS

A12.3.1 General comments

The entry of contaminants into an ecosystem can cause direct harm to organisms, or may indirectly affect their ability to survive or reproduce. Effects may be immediately apparent or become noticeable only after considerable delay. The effects on ecosystems are due in part to the physical and chemical properties of the contaminants themselves, but they also depend on the unique combination of physical, chemical and biological processes occurring in each ecosystem. In addition, exposed populations differ in their natural tolerance to particular contaminants, their behavioural and life-history characteristics, the dose to which they are exposed, and the exposure time. The time taken for exposed populations to recover from exposure may also vary.

Ecological risk assessment seeks to determine the nature, magnitude, and transience or permanence of observed or expected effects. This must be done in an environment that is itself changing and causing change in the organisms and systems under study. One critical goal of ecological risk assessment is to reduce the uncertainty associated with predicting and measuring adverse effects of contamination.

The main effects of contaminants are:

- reductions in population size
- changes in community structure
- changes in the structure and functions of ecosystems.

A flow diagram of possible effects following release of toxic pollutants into a river is given in Figure A12.3.

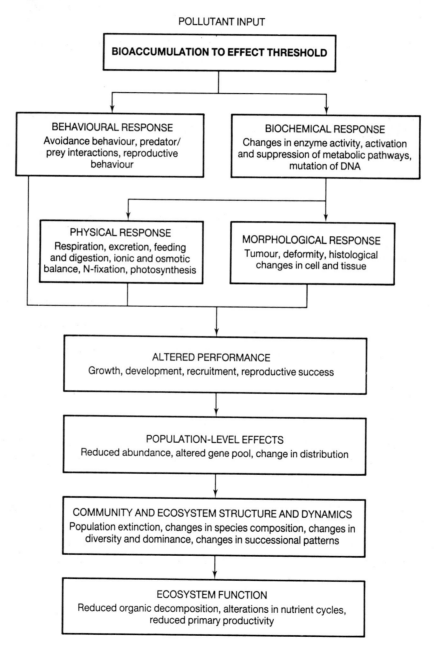

Figure A12.3 *River water quality: flow-diagram of induced effects following exposure to toxic pollutants (after Sheehan et al)*[10]

A12.3.2 Reduction in population size

Populations change in size through births, deaths, immigration and emigration. Contaminants can reduce population numbers through a range of mechanisms affecting one or more of these processes (see Section A12.3.4 and Box A12.4)

Box A12.4 *Causes of population decline due to contamination*

Changes in mortality rates caused by:

- exposure of some organisms to lethal doses
- depletion of a food source, perhaps by exposure to a contaminant, or because the contaminant allows a tolerant species to out compete other species for scarce resources
- exposure to sub-lethal doses which increases predation rates, or vulnerability to environmental changes.

Birth rates can decline due to:

- toxic effects
- through reduction of suitable breeding habitat
- changes in availability of high-quality food

Emigration and immigration may occur if :

- organisms sense and avoid contaminants
- sub-lethal effects cause changes in migratory behaviour.

A12.3.3 Changes in community structure

Many communities are constantly changing as a result of natural factors, for example :

- populations may increase and decrease seasonally and over longer periods
- predation and competition among species may bring about changes in the relative abundance of various species
- chance events, such as severe storms, may cause sudden increases in mortality of some species and open up habitats for others to colonise.

In the absence of a major disruption, changes in species composition and relative abundance in a community can be expected to occur relatively slowly and to vary within definable boundaries, perhaps cyclically or randomly.

The introduction of contaminants can introduce new boundaries, changing the range of possibilities in a way that is not always predictable, due to direct and indirect effects on the various populations making up the community. The effects of most concern are direct toxicity, changes in the environment, and changes in community structure brought about by changes in the population size of key organisms (see Box A12.5).

A12.3.4 Changes in ecosystem structure and function

As contaminants modify the species composition and relative abundance of populations in a community, the often complex patterns of matter and energy flow within the ecosystem also change. The elimination or reduction in numbers of a key species directly may interrupt the flow of energy and nutrients to other species not adversely affected directly by toxicity. For example:

- if plant life is adversely affected by a contaminant, the ecosystem as whole may capture less solar energy and thus support less animal life

- if microbial or invertebrate populations are disrupted, decomposition of dead plants and animals may not occur rapidly enough to supply sufficient mineral nutrients to sustain the plant community.

> **Box A12.5** *Effects of contaminants on community structure*
>
> **Direct toxicity**
>
> Direct toxic effects eliminate some species, reduce the numbers of some, and eliminate others. The relative proportions of species in these categories depends on the exposure level and the sensitivity of the species concerned.
>
> **Changes in the environment**
>
> Changes in the environment may have indirect effects, for example, a change in dissolved oxygen or salinity of an aquatic system may eliminate some species and favour others, creating an entirely new species mix and food web.
>
> **Changes in the size of particular populations**
>
> Changes in the size of particular populations may cause secondary changes in the composition of communities; for example, certain species may be a major source of food, provide shelter for the rest of the community, or be crucial in maintaining a balance of species in a habitat (e.g. if a predatory species is reduced or eliminated, the relative abundance of prey species may change significantly).

A12.4 FACTORS INFLUENCING THE ECOLOGICAL EFFECTS OF CONTAMINANTS

A12.4.1 General comments

Particular aspects of concern are:

- the nature of contamination as defined its physical, chemical and toxic properties and the frequency rate, and nature of its release to the environment

- the physical and chemical characteristics of the environment

- a range of biological factors including susceptibility of species, characteristics governing population abundance and distribution, temporal variability in communities and the movement of chemicals in food chains.

A12.4.2 Nature of contamination

Physical and chemical properties

Physical and chemical properties govern the transport and fate of chemicals in the environment, including their partitioning between environmental compartments and their tendency to bio-accumulate. Persistence (i.e. resistance to degradation) in different media and environmental compartments is an important characteristic.

Toxicity

The toxic properties of contaminants can significantly increase the mortality rate of populations by direct effects, or by changing an organism's ability to survive and reproduce in less direct ways, such as:

- altering developmental rates, metabolic processes, physiologic functions, or behaviour patterns
- increasing susceptibility to disease, parasitism, or predation
- disrupting reproductive functions
- causing mutations or otherwise reducing the viability of offspring.

Toxicity to humans is usually viewed in terms of the hazard presented, acute and chronic toxicity and dose-response relationships (see Section 5). The establishment of such basic information for complex ecosystems is extremely difficult and cannot be covered here (but see references 8 and 9).

A12.4.3 Nature of the release

Toxic chemicals may enter the environment, or move among environmental compartments, according to several possible timescales and in different carrier media (e.g. water, air, non-aqueous liquid etc.). For example, entry into a water course may occur:

- only once (e.g. an accidental spill)
- intermittently (e.g. storm water run-off)
- seasonally
- regularly (e.g. from daily on-site activities)
- continuously (e.g. groundwater seepage).

The effects of single or occasional releases are likely to be considerably different from those of a continuous release. Frequent releases of a non-persistent compound may have a long-term effect equivalent to a single release of a very persistent chemical. Occasional release may temporarily depress a population, but continuous release may trigger major changes in the composition of an ecosystem.

Different species of plants and animals may have different abilities to withstand or resist intermittent or continuous releases of toxic chemicals. For example, adults of a species may withstand a short-term discharge that kills all juveniles, but be severely affected by regular or continuous release. Similarly, chronic discharges that allow bio-accumulation of certain toxic chemicals may cause more lasting damage to certain species than to others.

A12.4.4 Physical/chemical characteristics of the environment

A wide range of environmental variables can influence both the nature and extent of the effect(s) of a contaminant on a living system. By interacting with each other, with contaminants, and with organisms, these factors can affect the outcome of contamination by:

- chemically changing the contaminant to make it less toxic
- making the contaminant more or less available in the environment
- making the organisms more or less tolerant of the chemical.

Important physical/chemical characteristics include temperature; pH; salinity of water; hardness of water, and soil composition.

A12.4.5 Biological factors

Biological factors governing the impact of contaminants include those listed in Box A12.6.

A12.5 PLANNING AN ECOLOGICAL ASSESSMENT

A12.5.1 Introduction

Problem formulation is the first phase of the ecological risk assessment which establishes the objectives, breadth and focus of the assessment. It should include:

- the general approach to be used to ensure the assessment provides the correct information, in a form which enables decisions to be made about appropriate risk reduction measures
- identification of stressor (contaminant) characteristics, ecosystems potentially at risk, and anticipated effects
- selection of appropriate assessment and measurement endpoints.

Problem formulation should lead to the development of a conceptual model embracing a series of hypotheses on how the contaminants might affect the ecosystem(s) potentially at risk, and the relationship between measurement and assessment endpoints.

Problem formulation should be based on a preliminary investigation. The principal aims of this should be to:

- determine if enough evidence exists to warrant investigation of ecological effects
- establish the scope of the ecological assessment (if one is judged necessary) in terms of spatial and temporal extent, tests to be conducted, time and resources needed, and level of detail required
- define study goals and data quality objectives if collection of new data is necessary.

Box A12.6 *Biological factors governing the effects of contaminants on ecosystems*

Susceptibility of species

Species differ in the way they take in, accumulate, metabolise and excrete contaminants, resulting in marked differences in susceptibility to particular contaminants. Response will depend mainly on dose but also on the rate of delivery, and the mode of exposure. In general, the susceptibility of a species depends on:

- the rate at which the contaminant is absorbed
- the resultant dose actually incurred at the physiological site where toxic effects occur within the organism
- sensitivity to the dose
- the relationship between the site of action and the expression of the symptoms of toxic injury
- the rate of repair or accommodation to the toxic injury.

Characteristics governing population abundance and distribution

For a given set of environmental conditions, species have characteristic attributes such as birth rates, and sex distribution, migrations patterns, and mortality rates. Habitat and food preferences, and other behavioural characteristics (e.g. nesting, foraging, rearing young) may also determine population size and distribution, and may significantly affect potential exposure.

Differences in response to contamination due to such characteristics may be evident :

- immediately e.g. a species with a high proportion of juveniles may decline more rapidly than one that has a higher proportion of adults with lower susceptibilities
- only after considerable time, when the stress has been removed from the environment, e.g. when the presence of contamination has favoured one species over another.

Pioneer species with high reproduction and dispersal rates are able to recolonise more rapidly than members of communities with lower reproduction rates, long generation times and longer individual lifetimes.

Temporal variability in communities

The effects of a contaminant discharge into a particular habitat may vary with seasonal or longer cycles governing community structure and function. Effects may be apparent immediately at one point in the cycle (e.g. in spring), but delayed at another time of year. Contaminants can also elicit different responses at different stages in a community's development. Seasonal changes are fairly predictable but longer-term successional changes are less regular and consequently predictable. Human intervention must also be taken into account.

Movement of chemicals in food chains

The processes involved in transfer and accumulation of chemicals via food webs are complex. Important points to take into account when assessing actual or potential transfers include:

- elevated concentrations of contaminants in organisms (compared to environmental concentrations) may not always be a sign of food-chain transfer as many organisms can bioaccumulate through direct exposure to soil, water, etc
- certain species are more exposed to food-chain accumulation, particularly those near the top of food chains (e.g. predators), or those which are long-lived, fattier or larger
- certain chemicals are more likely to be transferred than others
- plant foliage can become contaminated from soil by sorption of volatilised chemical on the leaves or by deposits of dust, aerosols, and vapours
- longer food chains increase the time needed to reach equilibrium levels in organisms (e.g. predators) at the top of the chain
- bio-accumulation may be less than predicted because of avoidance of contaminated prey or because of insufficient time to achieve equilibrium in living tissues
- very low bio-concentration factors can be significant.

The planning of the ecological risk assessment should take into account:

- the objectives of the assessment
- the time scales within which the assessment must be made
- the nature and quantities of contamination on the site
- the means of potential or actual release of contaminants
- the topography, hydrology, geology and other physical and spatial features of the site
- habitats potentially affected by the site
- the populations potentially exposed
- exposure pathways to potentially sensitive populations
- the possible or actual ecological effects of the contaminants or remedial actions.

A12.5.2 Determining need, objectives and level of effort

The need for an assessment is determined initially on the basis of available information (e.g. the results of the preliminary investigation) using specialist advice where necessary.

Typical objectives are to:

- document actual or potential threat of damage to the environment
- determine the extent of contamination
- determine the actual or potential effects of contaminants on protected wildlife species, habitats or special environments
- develop remediation criteria
- evaluate the potential ecological impacts of different remediation strategies.

A12.5.3 Evaluation of site characteristics

It is important to determine the full extent of the area that may be affected by contamination. This should start with a preliminary investigation using maps etc. and site visits to identify habitats of concern and their extent, potential pathways for contaminants movement etc. The size of the study should be defined by the potential for exposure, not by arbitrary distances or boundaries that lack biological justification.

For each habitat, sampling and analytical strategies should be prepared taking into account:

- required detection levels
- environmental media to be sampled and analysed
- the 'sensitivity' of the habitat
- 'toxicity tests' to be performed and species to be tested
- ecological (population, community, or ecosystem) effects to be measured or predicted.

The importance of habitats will vary depending on such factors as:

- the species native to the area and their significance
- the availability and quality of substitute habitats
- land use and management patterns in the area

- the value (economic, recreational, aesthetic etc.) placed on such habitats by local people and others.

Sensitive environments, which are unique or unusual or necessary for the propagation of key species, should be given particular attention.

The significance or uniqueness of an environment is often a subjective judgement that may be determined by social, aesthetic, or economic considerations. Some, such as critical habitats for endangered species are defined by law (e.g. SSSIs). Generally environments may be considered significant because they:

- are unusually small or large
- contain an unusually large number of species
- are extremely productive (e.g. an important fishery)
- contain species considered rare in the area
- are especially sensitive to disturbance.

A12.5.4 Contaminant evaluation

The sampling strategy should take account of existing information on the contaminants present, their physical and chemical properties, and their behaviour in biological systems (e.g. transfer through food chains). This is a far more complex task than a conventional site investigation for hazard assessment as described elsewhere in this Volume, and should only be undertaken by experts.

It may be necessary to assess the toxicity of chemical mixtures using a variety of laboratory and field tests or observations[2]. Toxicity tests (bioassays) may be applied to various media (water, soil, sediments) and life-forms (fish, invertebrates and plants). Such media tests have the benefit of integrating the effects of contaminant mixtures in the ambient media. However, their short duration and use of a small number of standard test species make them less than perfect predictors of effects on populations and ecosystems: they do not indicate what components caused the toxicity, and they cannot be readily used for predictive assessments[5]. The particular problems of toxicity testing of sediments have been discussed by Burton and Scott[7].

A12.5.5 Potential for exposure

Before the effects of a contaminant on an organism can be evaluated it is necessary to know its biological availability (i.e. how much of the chemical is actually or potentially reaching the point of exposure). This depends on the characteristics of the contaminant, the organism, and the environment, and may involve measurements quite different from those used for human health risk assessments. Examples include measurement of dissolved phase versus total metals in water and of whole body concentrations in forage fish versus fillet concentrations in commercial or game fish.

Exposure assessment should:

- identify organisms actually or potentially exposed
- identify significant routes of exposure
- determine the amounts and concentrations of each contaminant to which organisms are actually or potentially exposed
- determine the duration of each exposure

- determine the actual and potential frequency of exposure
- identify seasonal and climatic variations in conditions likely to affect exposure
- identify the site-specific geophysical, physical and chemical conditions affecting exposure.

Analysis of contaminants in tissues of exposed organisms can provide a link between environmental concentrations and the amounts of contaminants likely to reach the 'site of action' in the organism. However, up-take by living organisms is a very complex process and differs markedly between species or even variants (e.g. races, cultivars) of the same species. Sampling strategies and evaluation of the results must take this into account.

A12.5.6 Selection of assessment and measurement endpoints

Appropriate endpoints for the assessment must be selected based on the available information concerning the site, the contaminants and likely exposure pathways. These will include assessment endpoints that will drive the decision making process and measurement endpoints which are used in the field to approximate, represent, or lead to the assessment endpoint (see Box A12.7 for examples).

> **Box A12.7** *Examples of assessment and measurement endpoints*
>
> - A decline in a sport fish population (the assessment endpoint) may be evaluated using laboratory studies on the mortality of surrogate species, such as fathead minnow (the measurement endpoint)[3].
> - A reduction in the numbers of key aquatic species (the assessment endpoint) may be measured using water quality criteria for chronic toxicity (measurement endpoints).
> - The effects of a contaminant on a chalk grassland community may be assessed by monitoring species composition and frequency (measurement endpoints) with maintenance of a 'natural' assemblage providing the assessment endpoint.

As indicated above, the toxicity of contaminants to individual organisms can have consequences for populations, communities, and ecosystems. The effects occur through direct toxicity and through a number of interacting indirect routes. The effects can be spread over a considerable time period.

To characterise the effects of contaminants on populations, communities and ecosystems, one or more measures must be chosen depending on the objectives of the assessment.

Use of these measures will usually require comparison of the site to a carefully selected reference area or areas. This should :

- be close to the contaminated area(s)
- be closely resemble the area(s) of concern in terms of topography, soil composition, water chemistry, etc.
- have no apparent exposure pathways from the site in question or from other sources of contamination.

Measures that might be used to compare reference and contaminated areas are listed in Box A12.8. Suter and Loar[5] describe the use, and some of the difficulties encountered, of reference sites for studies of water courses in a catchment area with multiple sources of contamination.

> **Box A12.8** *Measures to be used to compare areas under investigation with reference areas*
>
> - population abundance
> - age structure
> - reproductive potential and fecundity
> - species diversity
> - food web or trophic diversity
> - nutrient retention or loss
> - standing crop or standing stock (the amount of biomass in the area)
> - productivity

When evaluating potentially affected habitats it is important to keep in mind that they may be affected by:

- direct or indirect exposure to contaminants
- physical disruption due to the design or operation of the site
- chemical disruption of ecosystem processes due to interference by contaminants with natural biochemical, physiological and behaviourial processes
- physical or chemical disturbance or destruction due to remedial activities
- other stresses not related to the site or its contaminants, such as extreme weather conditions or air pollution.

The evaluation of potentially affected populations will need to address such issues as:

- productivity and abundance
- the presence of rare, threatened, or endangered species
- potentially affected sport or commercial species.

A12.5.7 Sampling and analysis plan

The final step in the planning stage is to devise a sampling and analysis plan that will provide the data necessary to answer the various questions raised above. It must:

- meet the specific objectives of the sampling effort
- provide sufficient data (types, number, distribution, and timing)
- provide data of the required quality (a formal quality assurance plan should be prepared)
- cover all important pathways and environmental targets
- make good use of pre-existing data and sampling locations
- permit integration of sampling from different media and locations to allow integration of data (e.g. movement of contaminants along pathways).

The field sampling plan should take account of:

- actual or potential sources of contaminant release
- the media to which the contaminants can be or are being released
- the organisms that come into contact with contaminants

- the environmental conditions under which the transport and/or exposure may be taking place.

Identification of exposure routes and media should define the most appropriate plant and animal species to be sampled for analysis to determine contaminant concentrations, toxicity testing, or other measures of potential effects (e.g morphological examination). If food chain transfer of contaminants is suspected information on the trophic structures of affected ecosystems will be needed to determine which species should be examined for chemical residues.

Biological data collected in conjunction with these analyses may include parameters such as dry weight of tissues or organisms, percent moisture, lipid content, and the size, age or life stage of the organism. Contaminant concentrations may have to be expressed relative to the whole body weight or weight of the edible portion. Biological assessment techniques are described in more detail in Box A12.9.

Box A12.9 *Bioassessment techniques*

Bioassays

Laboratory and in-situ bioassays determine if contaminated media such as surface water, sediment, or soil are toxic to test organisms (e.g. fish, invertebrates, plants). These data can be obtained at relatively low cost and can be used to evaluate potential ecological impacts. However, the results are dependent on the specific test conditions such as test species, life stages, and exposure regimes. They do not provide a direct measure of the effects on resident biota or on their populations or communities.

Exposure biomarkers

Exposure biomarkers are physiological, biochemical, or chemical indicators of contaminant exposure used to measure individual exposure of organisms comprising the populations and communities at the site. Biomarkers are generally inexpensive to measure are broadly applicable across taxa and can be linked to broad classes of contaminants. They are not available for most potential contaminants. They provide a measure of exposure but seldom of toxic effect[9].

Biological surveys

Biological surveys measure the structural and functional characteristics of populations and communities. They are used to compare the biota at the site with biota at uncontaminated reference sites. Biological surveys provide an integrated measure of the actual effects in populations and communities but are generally expensive, time consuming, and often inconclusive[9].

Depending on the media to be sampled, the contaminants of concern, and the organisms under study, the sampling plan will also require collection of data on environmental conditions at the time of the study. Those required for aquatic environments are set out in Box A12.10.

For studies of potentially contaminated soil, information is needed on such parameters as particle size distribution, permeability and porosity, fraction and total organic carbon, pH, redox potential, cation exchange capacity, water content, colour, organic matter content, nutrient status and soil type. Guidance on appropriate descriptive/ analytical terms is provided in International Standards[10] and comparable national standards.

> **Box A12.10** *Examples of measurements to be made on aquatic environments*
>
> **Water quality**
>
> Hardness, pH, dissolved oxygen, salinity, temperature, presence or absence of thermocline, colour, dissolved organic carbon, and total suspended solids
>
> **Hydrological characteristics**
>
> Flow rate, groundwater discharge/recharge rates, aquifer thickness and hydraulic conductivity, depth, velocity and direction of current, tidal cycle and heights, surface water inputs and outputs
>
> **Sediment characteristics**
>
> Grain size distribution, permeability and porosity, bulk density, organic carbon content, pH, colour, mineralogical composition, benthic oxygen conditions and water content.

A12.6 ESTIMATION AND EVALUATION OF RISKS

The evaluation of risks to ecological targets associated with contaminated sites should address:

- the probability that an adverse effect will occur
- the magnitude of each effect
- the temporal character of each effect (transient, reversible, permanent)
- the target populations or habitats likely to be affected.

Depending on the objectives of the assessment and the quality of the data collected, the answers to these questions will be expressed qualitatively, quantitatively, or semi-quantitatively. The same principles of hazard and risk assessment described in Section 5 should apply.

Exceedance of any accepted criteria/guidelines for individual species or media (e.g. water quality standards) provides one measure (the source or derivation of any guidelines or criteria used should be clearly stated).

However, beyond this a large measure of professional judgement will be required to complete ecological risk estimation and evaluation. Questions concerning spatial and temporal components and the possible impacts of remedial responses, such as those listed in Box A12.11, are most likely to be answerable only in qualitative terms, as statements of the best judgement of the ecologists involved.

> **Box A12.11** *Temporal, spatial and other considerations in ecological risk characterisation*
>
> How long will the effects last if the contaminants are removed?
>
> How long will it take for receptor population to recover?
>
> Will there be inter-generational effects?
>
> Will the contaminants move beyond the area of study through biotic transport?
>
> What effect will remediation have on this process?
>
> If there are community and ecosystem effects of contamination, is removal of the contaminants sufficient to restore community structure and ecosystem functioning?
>
> If not what else will be needed?
>
> How do the data on exposure and observed or predicted effects relate to the rapidity of response required?
>
> Which responses are required immediately?
>
> Which responses can or should be undertaken later?
>
> What limits will the proposed remediation strategy or mitigation actions place on future options for further remediation, follow-up assessment, and resource use?

Suter and Loar[5] describe the process of the evaluation as one of ecological epidemiology and liken it to civil court proceedings in which the balance of probability is decided on the basis of a variety of evidential sources (see Box A12.12 and Figure A12.4).

It is important to remember that many populations and ecosystems exhibit considerable resilience in the face of disturbance. Change is more common in ecosystems than stability. In many situations, when a source of contamination is removed, natural systems will rapidly recover their former appearance. Hence, for the same amount of chemical released, the risk associated with an acutely toxic but short-lived chemical may be considered important but less so than a moderately toxic, highly persistent, chemical.

An example of an aquatic risk assessment is provided in Box A12.13. For other examples see references 4, 5, and 11.

> **Box A12.12** *Ecological epidemiology*[5]
>
> In ecological epidemiology,... 'the traditional risk paradigm of inferring effects from source information is only one line of evidence. In addition, the nature and distribution of measured effects are used to make inferences about the significance of sources, and evidence of exposure is used to infer how sources are contributing to the exposure experienced by organisms and which effects can be credibly attributed to contaminants (Figure A12.4). In this ecological epidemiology, risk is not the result of a single line of evidence running from source to risk, but a braid of lines of evidence. It resembles a civil court case compounded of diverse expert testimony, witness testimony, physical evidence, and forensic evidence, and in which the verdict is based on the preponderance of evidence.'

Box A12.13 *An example of an aquatic ecological risk assessment*

Cardwell *et al*[12] have described a three-tier aquatic ecological methodology and the application of the first two tiers to determine the risk to aquatic life from acid mine drainage. It is based on the assumption that sufficient good quality data have been obtained over a period of time to enable probabilistic profiles of concentrations against time of each contaminant of concern to be established.

The three tiers are:

1. Screening-level risk assessment
2. Risk quantification with existing data
3. Risk quantification with new data.

The three tiers vary in reliability, cost, completion time, and data requirements. Substitution of site-specific data and reliance on fewer simplifying assumptions distinguish Tier 3 from Tier 2. Tier 1 produces valid but more conservative (worst case) estimates of risk than Tiers 2 and 3.

In the case described, Tier 1 and 2 assessments were used to evaluate acid mine drainage surfacing to form a stream confluent with a scenic river inhabited by a threatened fish species, the round tail chub (*Gila robusta*). The acid mine drainage has undergone natural pH neutralisation in the aquifer and only the most mobile metals persist in the stream. The contaminants of potential concern were: aluminium, antimony, barium, boron, cadmium, chromium, cobalt, copper, iron, manganese, nickel and zinc.

The assessment endpoint was maintenance of a balanced, indigenous community of aquatic life.

For the Tier 1 assessment, the measurement endpoints were the US Environmental Protection Agency's water quality criteria for chronic toxicity.

In the Tier 1 screening process, 95% upper confidence limits for expected environmental concentrations (EECs) were compared with the EPA water quality criteria for chronic toxicity (CV). Those for which the hazard quotient (EEC/CV) exceeded 0.3 were selected for assessment in the Tier 2 process. Those selected were: aluminium, cadmium, cobalt and manganese. Although chemicals with hazard quotients (HQs) >1.0 are those which by definition pose a potential risk, those with HQs >0.3 were included in the Tier 2 process because they may contribute to chronic effects resulting from additivity or synergism.

In the Tier 2 process, the expected total risk is calculated from probabilistic distributions of EECs and the % of the taxa identified as likely to be affected by particular concentrations of each contaminant derived from published sources and laboratory studies. For example, an EEC for aluminium of 16.8 μg/litre occurs 15.4% of the time and is expected to affect 1.8% of the aquatic community taxa. The net risk is 0.28%. The total risk is obtained by summing the net risks for different values of the EEC. The total risk for aluminium was calculated to be 6.43% i.e. 6.43% of species are expected to be affected by the measured aluminium concentrations. In contrast 100% of the aquatic species were predicted to be at chronic risk from manganese and 80% of acute toxicity at the median concentration.

The risk assessment predicted chronic toxicity to the majority of the aquatic species. A biosurvey suggested a 70% overall reduction in taxa.

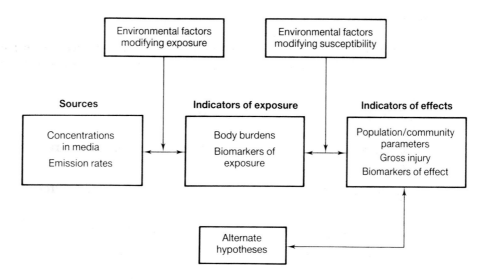

Figure A12.4 *Diagram of risk characterisation in ecological epidemiology*

A12.7 REPORTING THE RESULTS

A12.7.1 Introduction

The results of the assessment should be presented in a report or series of reports and include:

- a statement of the objectives
- the scope of the study
- a description of the site and surrounding area studied
- a description of the contaminants present
- exposure evaluation
- evaluation of hazards and risks
- derivation of any guidelines or criteria used
- a description of the limitations of the assessment
- conclusions
- recommendations.

A12.7.2 Objectives of the assessment

The objectives of the assessment should be clearly stated together with a statement on whether pre-existing or on newly acquired data have been used. The general approach adopted should be :

- assess current effects
- assess past damage
- predict future effects
- evaluate a combination of these.

A12.7.3 Scope

This should describe the amount and type of information collected, the methods use of the time frame of the study (i.e. the time period and seasons in which the data were collected), and the sampling intervals.

A12.7.4 Description of site and area

Descriptions should be provided supported by maps, plans, photographs and relevant data including:

- the site and surrounding habitats in physical, topographical, hydrological, geological and ecological terms
- any obvious signs of contamination (see also Box A12.10)
- likely exposure pathways
- any obvious effects potentially attributable to contamination.

A12.7.5 Description of contaminants

The compounds of ecological concern and the reasons why they are of concern should be identified.

The results of all chemical analyses and other measurements should be presented together with such statistical information as is necessary to evaluate them and to establish their validity.

A12.7.6 Exposure evaluation

Actual and potential pathways should be identified, taking into account environmental fate and transport through physical and biological means. The report should describe each pathway by chemical(s) and media involved, and identify the pathways in space and time with respect to the site and the period of investigation. Models used in the exposure assessment, and any assumptions used should be described in full. Data should be presented to show clearly temporal and spatial variations.

A12.7.7 Estimation and evaluation of hazards and risks

The evaluation of hazards, and estimation and evaluation of risks should take account of the discussion in Section A12.6. The risk estimation methods employed, the uncertainties surrounding them, and the wider societal and other factors taken into account in assessing the significance of the risks should be clearly stated.

A12.7.8 Conclusions and limitations of assessment

The process of assessing ecological effects is one of estimation under conditions of uncertainty. It is essential therefore that all data are reported with appropriate statistical information on the degree of confidence attached, and that all uncertainties and assumptions are stated in full.

Ecological assessment is, and will remain, a process combining careful observation, data collection, testing and professional judgement. By carefully describing the sources of uncertainty, confidence in the conclusions reached can be strengthened.

A12.8 RESPONSES TO ECOLOGICAL RISK ASSESSMENTS

The response to the risk evaluation in cases where damage has occurred will depend on many factors including:

- the value of the ecosystem
- the practicability of repairing the damage
- the cost of repairing the damage.

Where the damage is threatened by proposed remedial action the response will depend on:

- the value of the ecosystem
- the possibilities for avoiding damage through protective measures or a change in the remedial strategy
- the cost of providing protective measures or developing an alternative remediation strategy.

REFERENCES

1. US ENVIRONMENTAL PROTECTION AGENCY. *Risk assessment guidance for Superfund, Volume II: Environmental evaluation manual.* US Environmental Protection Agency, Office of Emergency and Remedial Response, Washington DC 1989, EPA/540/1−89/001.

2. US ENVIRONMENTAL PROTECTION AGENCY. *Ecological assessments of hazardous waste sites: A field and laboratory reference document.* US Environmental Protection Agency, Office of Emergency and Remedial Response, Washington DC 1989, EPA/600/3−89/013.

3. US ENVIRONMENTAL PROTECTION AGENCY. *Framework for ecological risk assessment.* US Environmental Protection Agency, Risk Assessment Forum (Washington DC) 1992, EPA/630/R-92/001.

4. WATER ENVIRONMENT FEDERATION. *Application of ecological risk assessment to hazardous waste site remediation.* Water Environment Federation (Alexandria VA, USA) 1993.

5. SUTER, G.W. and LOAR, J.M. Weighing the ecological risk of hazardous waste sites. *Environmental Science and Technology*, 1992, **26**(3), 432-438.

6. ADAMS, W.J., KIMERLE, R.A. and BARNETT, J.W. Sediment quality and aquatic life assessment. *Environmental Science and Technology*, 1992, **26**(10), 1864-1875.

7. BURTON, G.A. and SCOTT, K.J. Sediment toxicity evaluations: their niche in ecological assessments. *Environmental Science and Technology*, 1992, **26**(11), 2068-2075.

8. SUTER, G.W. (ed.) *Ecological Risk Assessment.* Lewis Publishers (BOCA Raton, Florida) 1993

9. MAUGHAN, J.T. *Ecological Assessment of Hazardous Waste Sites.* Van Nostrand Reinhold (New York), 1993

10. SHEEHAN, P.J., MILLER, D.R., BUTLER, G.C., and BORDEAU Ph (ed.) *Effects of pollutants at the ecosystem level.* Wiley (Chichester), 1984

11. DURDA, J.L. Ecological risk assessments under Superfund. *Water Environment and Technology* 1993, **5**(4), 42-46.

12. INTERNATIONAL ORGANISATION FOR STANDARDISATION. *Soil quality – Description of soils and sites.* Committee Draft CD 11259 (obtainable from Secretary, Committee EPC 48, British Standards Institution (London).

13. HUGGETT, R.J., UNGER, M.A., SELIGMAN, P.F. and VALKIRS, A.O. The marine biocide tributyltin: assessing and managing the environmental risks. *Environmental Science and Technology*, 1992, **26**(2), 232-237.

14. CARDWELL, R.D., PARKHURST, B.R., WARREN-HICKS, W. and VOLOSIN, J.S. Aquatic ecological risk. *Water Environment and Technology* 1993, **5**(4), 47-51.

Further reading

McCARTHY, L.S. and MacKAY, D. Enhancing ecotoxicological modelling and assessment. *Environmental Science and Technology*, 1993, **27**, (9), 1714-1725